Springer Theses

Recognizing Outstanding Ph.D. Research

Aims and Scope

The series "Springer Theses" brings together a selection of the very best Ph.D. theses from around the world and across the physical sciences. Nominated and endorsed by two recognized specialists, each published volume has been selected for its scientific excellence and the high impact of its contents for the pertinent field of research. For greater accessibility to non-specialists, the published versions include an extended introduction, as well as a foreword by the student's supervisor explaining the special relevance of the work for the field. As a whole, the series will provide a valuable resource both for newcomers to the research fields described, and for other scientists seeking detailed background information on special questions. Finally, it provides an accredited documentation of the valuable contributions made by today's younger generation of scientists.

Theses are accepted into the series by invited nomination only and must fulfill all of the following criteria

- They must be written in good English.
- The topic should fall within the confines of Chemistry, Physics, Earth Sciences, Engineering and related interdisciplinary fields such as Materials, Nanoscience, Chemical Engineering, Complex Systems and Biophysics.
- The work reported in the thesis must represent a significant scientific advance.
- If the thesis includes previously published material, permission to reproduce this must be gained from the respective copyright holder.
- They must have been examined and passed during the 12 months prior to nomination.
- Each thesis should include a foreword by the supervisor outlining the significance of its content.
- The theses should have a clearly defined structure including an introduction accessible to scientists not expert in that particular field.

More information about this series at http://www.springer.com/series/8790

Yutaka Hirai

Understanding the Enrichment of Heavy Elements by the Chemodynamical Evolution Models of Dwarf Galaxies

Doctoral Thesis accepted by
the University of Tokyo, Tokyo, Japan

 Springer

Author
Dr. Yutaka Hirai
RIKEN Center for Computational Science
Hyogo, Japan

Supervisor
Prof. Toshitaka Kajino
National Astronomical Observatory
of Japan
Beihang University
The University of Tokyo
Tokyo, Japan

ISSN 2190-5053   ISSN 2190-5061   (electronic)
Springer Theses
ISBN 978-981-13-7883-6   ISBN 978-981-13-7884-3   (eBook)
https://doi.org/10.1007/978-981-13-7884-3

This Springer imprint is published by the registered company Springer Nature Singapore Pte Ltd.
The registered company address is: 152 Beach Road, #21-01/04 Gateway East, Singapore 189721, Singapore

Supervisor's Foreword

The discovery of radioactivity by Antoine Henri Becquerel in 1896 and the extraction of radioactive substances by Marie Curie in 1898 opened a new window to modern science at the beginning of the past century. The key radioactive atomic nuclides in these discoveries were ^{226}Ra (half-life $t_{1/2} = 1600$y), ^{232}Th ($t_{1/2} = 140.5$Gy), and ^{238}U ($t_{1/2} = 44.7$Gy) that are used today as cosmic clocks in astronomy of dating the age when astrophysical event producing these elements such as supernova explosion or binary neutron star merger happened in the history of the universe. Although these atomic nuclides are thought to be produced in the rapid neutron-capture process (abbreviated as r-process hereafter), the astrophysical site has still not been uniquely identified. This is one of the most fundamental unanswered questions in modern science of the twenty-first century, as referred among "The 11 Greatest Unanswered Questions of Physics" declared by The National Research Council's Board on Physics of the United States of America in 2002.

It is the recent focus in astronomy, too, to find a solution to this question because these heavy atomic nuclides are the invaluable "observable tracer" of the galactic evolution from the early universe until the recent epoch of the solar system formation along the long history of cosmic evolution. Although the leading candidate site of neutrino-heated supernova explosion had been put into question, magneto-hydrodynamic jet supernova was proposed as an alternative site for the r-process. The recent discovery of a binary neutron star merger GW170817/SSS17a also has raised a hope, but no emission lines from the r-process elements were detected. Nevertheless, this was an event of the century that opened a new window to multi-messenger astronomy and astrophysics. Optical and near-infrared emissions along with the detection of gravitational wave and gamma-ray bursts strongly suggest that the r-process occurred because their total energy release is consistent with radiative decays of theoretical prediction of the r-process nuclei.

The Ph.D. thesis of Yutaka Hirai at the University of Tokyo, who is currently a research associate at RIKEN Center for Computational Science in Japan, was written in this background.

Yutaka Hirai performed a series of high-resolution simulations of the chemo-dynamical evolution of dwarf spheroidal galaxies to understand the enrichment history of heavy elements by using supercomputers and others equipped at National Astronomical Observatory of Japan. The basic difference between neutron star merger and supernova is the emergent event rate as a function of cosmic time. Supernova can explode in a few Mys from the early galaxy, while binary neutron stars rarely merge at the rate of 0.1–1% of supernova rate, and also their coalescence time delays by narrowing orbital motion due to very slow gravitational wave radiation. It was, therefore, a general argument that the neutron star mergers could not contribute very much to the early galaxy. Yutaka Hirai carried out numerical N-body SPH simulations by taking account of gaseous hydrodynamics including stars and dark matters, coupled with star formation, supernova and merger explosions, and their feedback onto materials including atomic nuclides and heat. He then finds that the neutron star mergers can produce r-process elements even at the very low metallicity region due to the suppressed star formation in dwarf spheroidal galaxies, although the result is somewhat subject to several parameters in chemo-dynamical evolution model. This theoretical discovery sheds light on our understanding of the relation between star formation histories and enrichment of the r-process elements in metal-poor stars.

Yutaka Hirai also extensively studied the galactic chemo-dynamical evolution by taking account of different kinds of supernovae such as electron-capture supernovae, hypernovae, and type Ia supernovae in order to clarify the roles of each episode in the galactic chemical evolution. He finds that electron-capture supernovae can contribute to the formation of stars, which exhibit high Zn-to-Fe ratios as seen universally in astronomical observations of low metallicity stars.

A series of Yutaka Hirai's studies of galactic chemo-dynamical evolution are summarized in this book. His sophisticated, detailed hydrodynamic simulations and numerical techniques presented in this book have made remarkable progress in understanding the galaxy evolution in terms of abundance variation of heavy atomic nuclides in metal-poor stars. One of the highlights from these studies is that he finds the heavy elemental abundances as a unique and quantitative indicator to show the efficiency of metal mixing in the galaxy. Another novel finding is that the timescale of metal mixing shorter than the dynamical time results in the lack of r-process enhanced stars in dwarf spheroidal galaxies.

Many other new theoretical findings in Yutaka Hirai's works have made remarkable progress in studying the formation and evolution of galaxies in terms of the r-process elements in metal-poor stars. Ongoing and near future astronomical surveys of metal-poor stars are expected to increase the number of observations of the r-process elements. A close comparison between theoretical simulations and observations will certainly clarify more details of the evolutionary histories of the Local Group galaxies.

Tokyo, Japan   Toshitaka Kajino
April 2019

Parts of this thesis have been published in the following journal articles:

- Yutaka Hirai, Yuhri Ishimaru, Takayuki R. Saitoh, Michiko S. Fujii, Jun Hidaka, and Toshitaka Kajino 2015, *The Astrophysical Journal*, 814:41
- Yutaka Hirai, Yuhri Ishimaru, Takayuki R. Saitoh, Michiko S. Fujii, Jun Hidaka, and Toshitaka Kajino 2017, *Monthly Notices of the Royal Astronomical Society*, 466, 2474–2487
- Yutaka Hirai and Takayuki R. Saitoh 2017 *The Astrophysical Journal Letters*, 838:L23
- Yutaka Hirai, Takayuki R. Saitoh, Yuhri Ishimaru, and Shinya Wanajo 2018 *The Astrophysical Journal*, 855:63

Acknowledgements

I would like to express my sincere gratitude to Toshitaka Kajino for his fruitful advice, endless patience, and continuous encouragement. I am very grateful to him for encouraging me to study galactic chemodynamics. I also thank him for giving me many opportunities to present my research and introducing me to many outstanding researchers.

I am deeply grateful to Yuhri Ishimaru for continuous advice and fruitful discussion. Her works motivate me to do my research projects. I also thank her for giving me an opportunity to learn and discuss galactic chemical evolution with researchers and students at her laboratory. I would like to express my great pleasure working with her.

I am very grateful to Michiko Fujii for giving me suggestions to improve my work. I appreciate her very much for continuously giving me advice on the directions of my research project. I also thank her for introducing me and making connections to many researchers in astronomy.

I appreciate very much for Takayuki Saitoh for providing me the code, ASURA, and giving me advice for extending the code. I learned a lot about how to conduct large-scale numerical simulations from him. Fruitful discussion with him makes my works improve significantly.

I would like to thank Shinya Wanajo for fruitful discussions and providing me the nucleosynthesis yields. Collaboration with him makes us improve the understanding of the origin of elements in the Universe. I would like to thank Jun Hidaka for continuous advice from the beginning of this work. I appreciate the referees of this thesis, Hideyuki Umeda, Takashi Onaka, Eiichiro Kokubo, Masashi Chiba, and Wako Aoki.

I immensely appreciate Masaomi Tanaka for encouraging me to do this work. His advice motivates me to start new projects. I appreciate Takuma Suda for providing me the latest version of SAGA database including the dataset of Local Group dwarf galaxies. I am deeply grateful to Yuzuru Yoshii and Toshikazu Shigeyama for giving me insightful suggestions. I thank Ke-Jung Chen for giving me suggestions and introducing me Taiwanese astronomers. I appreciate Nozomu

Tominaga, Takashi Okamoto, Meng-Ru Wu, Yuichiro Sekiguchi, Nobuya Nishimura, and Aurora Simionescu for giving me insightful comments. I also appreciate Nobuyuki Hasebe and Hiroshi Nagaoka for encouraging me to proceed with research activities.

I have conducted my research at Division of Theoretical Astronomy (DTA), National Astronomical Observatory of Japan (NAOJ) as a student of the University of Tokyo. People in these institutes have supported my daily research activities. I sincerely appreciate Kohji Tomisaka for giving me suggestions and supporting my works in DTA. I am grateful to Tomoya Takiwaki, Takashi Moriya, Takaya Nozawa, Hajime Sotani, Jian Jiang, and Masaki Yamaguchi for giving me insightful comments in daily discussions. I would like to thank Shota Shibagaki, Akimasa Kataoka, Masato Shirasaki, Kenneth Wong, and Mariko Nomura for giving me advice for my works. I appreciate Shihoko Izumi, Hinako Fukushi, and Yuko Kimura for supporting my research activities. I thank all members I met in DTA: Haruo Yoshida, Fumitaka Nakamura, Ken Osuga, Takashi Hamana, Tsuyoshi Inoue, Seiji Zenitani, Dai Yamazaki, Kengo Tomida, Masahiro Ogihara, Hiroyuki Takahashi, Tomohisa Kawashima, Takayoshi Kusune, Benjamin Wu, Keizo Fujimoto, Yasuhiro Hasegawa, Shun Furusawa, Akihiro Suzuki, Yasunori Hori, Takami Kuroda, Ko Nakamura, Yukari Ohtani, Yuta Asahina, Hirokazu Sasaki, Kanji Mori, Yudong Luo, Shian Izumida, Hiroshi Kobayashi, Takashi Shibata, Misako Tatsuuma, Shuri Oyamada, Kangrou Guo, Akihiko Fujii, and Katsuya Hashizume. I am also grateful to visiting scientists at DTA: Grant J. Mathews, Roland Diehl, Cemsinan Deliduman, Yamaç Deliduman, Michael Famiano, A. Baha Balantekin, Shunji Nishimura, Tomoyuki Maruyama, Toshio Suzuki, and Takehito Hayakawa. Interactions with them have enhanced my works.

People in Particle Simulator Research Team and Co-Design Team at RIKEN Center for Computational Science (R-CCS) support me to continue my studies. I would like to express gratitude to Junichiro Makino for hosting me and giving me advice for future directions. I thank Masaki Iwasawa, Keigo Nitadori, Steven Rieder, Daisuke Namekata, Kentaro Nomura, and Yohei Ishihara for daily discussion. I am grateful to Yuko Wakamatsu, Miyuki Tsubouchi, and Yoshie Yamaguchi for supporting me to begin my life at R-CCS.

Seminars and discussions with people in many fields enhance my knowledge. I am grateful to Nicolas Prantzos, Patrick François, Simon Portegies Zwart, and Diederik Kruijssen for discussions during my short stay in Europe. Seminars with Keiichi Wada, Masahiro Nagashima, Motohiko Enoki, Shigeki Inoue, Taira Oogi, and Hikari Shirakata enhance my knowledge of galaxy formation. For galactic archaeology science, Miho Ishigaki, Kohei Hayashi, Hidetomo Homma, Daisuke Toyouchi, Takanobu Kirihara, and Tadafumi Matsuno gave me fruitful comments and suggestions. I appreciate Misa Aoki, Sachie Tsukamoto, Nao Fukagawa, and Takuya Ojima for studying together about galactic chemical evolution. I am grateful to Akiyuki Tokuno, Shalini Selvam, Asami Komada, and Padmavathi Jayajeevan for helping me edit the manuscript of this book.

Finally, I sincerely thank my family and friends for supporting my daily life. I am especially grateful to my mother, I.H., my father, S.H., and my grandparents. I appreciate S.S. for giving me great days.

This work was supported by JSPS KAKENHI Grant Numbers, 15J00548 18H05876, 19K21057, and 19H01933. The author is supported by the Special Postdoctoral Researchers (SPDR) program at RIKEN. This study was also supported by JSPS and CNRS under the Japan-France Research Cooperative Program. Numerical computations and analysis were in part carried out on Cray XC30, XC50, and computers at Center for Computational Astrophysics, National Astronomical Observatory of Japan and the Yukawa Institute Computer Facility. This work has made use of NASA's Astrophysics Data System.

Contents

Chapter 1
Introduction

Abstract Understanding the enrichment of heavy elements in galaxies helps us clarify the astrophysical sites of elements and chemodynamical evolution of galaxies. At the time of the big bang, the Universe consists of hydrogen, helium, and lithium. Explosive events such as supernovae and neutron star mergers distribute heavy elements to the interstellar medium. Dense gases containing these elements form stars. Abundances of heavy elements in stars, therefore, reflect the astrophysical sites of these elements and star formation histories in galaxies. Dwarf galaxies are the ideal sites to study the enrichment of heavy elements. According to the hierarchical structure formation scenario, clustering of smaller systems builds up galaxies. Dwarf galaxies would be survivors from the early stages of galaxy formation. In the Local Group, there are over 100 dwarf galaxies. These galaxies with resolved stellar populations help us understand the astrophysical sites of heavy elements and chemodynamical evolution of galaxies. In this book, we focus on the enrichment of elements synthesized by the rapid neutron-capture process (r-process elements) and zinc in dwarf galaxies. These elements show characteristic features in the chemical abundances of metal-poor stars. In this chapter, we review the astrophysical sites of elements, the observed Local Group galaxies, formation and evolution of galaxies.

1.1 Astrophysical Sites of Elements

1.1.1 Solar System Abundances

The solar system abundances give us information about the astrophysical sites of elements. We can determine the chemical compositions of the Sun by the absorption lines in the solar photosphere or meteorites in a group of the CI chondrite (e.g., [12]). Figure 1.1 shows the solar system isotopic abundances [83]. As shown in this figure, the abundances exponentially decline to mass number, $A \sim 100$ and constant after that. There is the strong peak at $A = 56$ because the ^{56}Fe nucleus is the most stable nucleus among all isotopes. The peaks at $A = 80, 130, 196$ and $A = 90, 138, 208$ are due to the the rapid neutron-capture process (r-process) and the slow neutron-capture

© Springer Nature Singapore Pte Ltd. 2019
Y. Hirai, *Understanding the Enrichment of Heavy Elements
by the Chemodynamical Evolution Models of Dwarf Galaxies*,
Springer Theses, https://doi.org/10.1007/978-981-13-7884-3_1

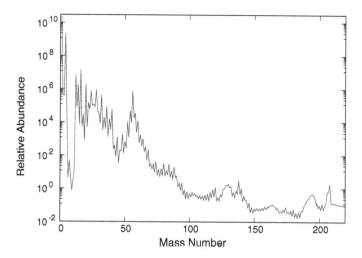

Fig. 1.1 The solar isotopic abundances relative to silicon (Si) abundance as a function of mass numbers [83]. The Si abundance is set to be 10^6

process (s-process), respectively. These peaks come from the magic number ($N = 50$, 82, and 126) of the nucleus.

The solar system abundance patterns of r-process elements match those of metal-poor stars (e.g., [136]). Figure 1.2 compares the abundance patterns of r-process elements in the Sun and metal-poor stars. According to this figure, elements from $50 \leq A < 90$ exactly match the solar system abundance pattern in metal-poor stars. Lighter and heavier r-process elements show small variations. This result indicates that nucleosynthetic events synthesize r-process elements from $50 \leq A < 90$ with the same pattern throughout the evolutionary history of galaxies. In this study, europium (Eu) is regarded as the representative of the r-process elements.

1.1.2 Nucleosynthesis

Synthesis of elements in the Universe has three paths: big bang, stellar nucleosynthesis, and destruction of a nucleus by a cosmic ray. Big-bang synthesizes elements from H to Li. No stable isotopes in the mass number of 5 and 6 prevent forming heavier elements except for ^{7}Li. Destruction of a nucleus by cosmic ray synthesize small kinds of elements such as Be and B.

Stars synthesize heavy elements. Burbidge et al. [22] and Cameron [24] systematically studied stellar nucleosynthesis. Major burning stages in stars synthesize elements from He to Fe. The proton-proton chain and CNO cycle in hydrogen burning convert four ^{1}H to ^{4}He (α). Later stages of stellar evolution have He, C, O, ... burning stages. These burning stages synthesize elements up to iron-peak elements.

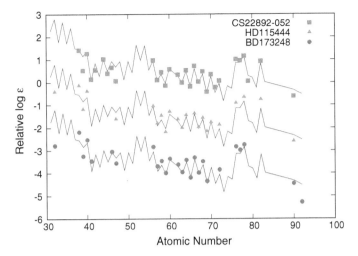

Fig. 1.2 Comparison of solar r-process abundance patterns with r-rich metal-poor stars. Green squares, orange triangles, blue circles represent observed abundances of metal-poor stars, CS22892-052, HD115444, and BD17 3248. Observed data are taken from Westin et al. [166], Cowan et al. [30], Sneden et al. [136, 137]. The solid purple curves represent the scaled solar r-process abundance pattern [131], normalized by the Eu abundance. The abundances except for CS 22892-052 is shifted for comparison

However, it is hard to synthesize isotopes heavier than ^{56}Fe because the binding energy of a nucleus is maximum around this mass number.

The capture of neutrons (n) mainly synthesize elements heavier than the iron-peak. The neutron-capture processes are divided into two processes. If the timescale of β-decay (t_β) is shorter than the timescale of neutron-capture (t_n), elements are synthesized along with the stable isotopes (s-process). The s-process cannot synthesize isotopes heavier than ^{209}Bi, which is the heaviest stable isotope.

The other neutron-capture process, the r-process occurs when $t_\beta \gg t_n$. Nucleosynthesis in the r-process occurs along with the nuclear limit lines. For the r-process, the neutron-rich environment is necessary. Kratz et al. [79] showed that the number density of neutron should be $\sim 10^{20}$–10^{30} cm^{-3} for the r-process to reproduce the solar abundances of r-process elements. The electron fraction (Y_e) indicates the proton-to-neutron ratio. If $Y_e < 0.5$, the number of neutrons is larger than that of protons. Wanajo [160] showed that the condition to synthesize elements beyond the third peak ($A \sim 196$) is $Y_e < 0.2$ if the asymptotic entropy (S) is less than 100 k_B nucleon^{-1}, where k_B is the Boltzmann constant.

High entropy environment is also suitable for the r-process. Since S is proportional to T^3/ρ, where T is the temperature, and ρ is the density, high entropy means that the density is low at a given temperature. This leads to be more difficult to synthesize seed nuclei through the reaction: $\alpha + \alpha + n \rightarrow {}^9$Be. As a result, the number of neutron per one seed nuclei becomes high. In the case of $S > 100$ k_B nucleon^{-1}, the condition to synthesize elements heavier than the third peak is

Fig. 1.3 The metallicity
dependence on final fates of
stars as a function of
zero-age sequence mass
(M_{ini}, Fig. 5 of Doherty et al.
[40] reproduced by
permission of the Oxford
University Press). The
dark-gray shaded, the
light-grey, white, hatched,
and gray regions denote the
mass range of CO white
dwarfs (WDs), CO(Ne)
WDs, ONe WDs,
electron-capture supernovae
(EC-SN), and iron
core-collapse supernovae
(CC-SN), respectively

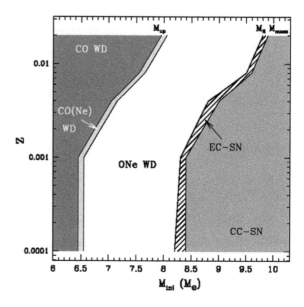

$(S/230 \, k_B \text{ nucleon}^{-1})/((Y_e/0.4)(\tau/20 \text{ ms})) > 1$, where τ is the expansion timescale
of the ejecta [160]. The astrophysical sites of the r-process should satisfy these conditions.

1.1.3 Astrophysical Sites of Heavy Elements

1.1.3.1 Supernovae

Massive stars explode as core-collapse supernovae at the end of their life. The final
fates of stars depend on the metallicity. Figure 1.3 shows final fates of stars. Stars
within a mass range between 8–10 $M_{\odot}$[1] develop oxygen-neon-magnesium core by
the carbon-burning. In these stars, low reaction thresholds of Ne and Mg increase
electron Fermi energy and electron capture triggers the gravitational collapse. These
stars explode as electron-capture supernovae (e.g., [60, 96, 97, 103, 104, 106]).
Wanajo et al. [161] estimated that the upper limit of the rate of electron-capture
supernovae is ∼4% of all core-collapse supernovae.

Stars more massive than 10 $M_{\odot}$ explode as iron core-collapse supernovae. When
the core of the star reaches nuclear statistical equilibrium at $T \sim 10^{10}$ K, the dis-
sociation of Fe-group nuclei to α particles and a growing number of free nucleons,
leading to the core collapse (e.g. [66]). The iron core-collapse supernovae are main
sources of heavy elements from carbon through zinc (e.g., [105, 170]).

---

[1] The symbol "$M_{\odot}$" means the solar mass ($=1.989 \times 10^{33}$ g).

Some fractions of stars more massive than 20 $M_\odot$ explode with kinetic energy ~1 dex higher than that of normal core-collapse supernovae. These supernovae are called hypernovae. Hypernovae are thought to be associated with long gamma-ray bursts. Observations of long gamma-ray bursts suggest that the rate of hypernovae is fewer than 1% of all type Ib/c supernovae [56, 114]. Hypernovae can be sources of iron-peak elements (e.g., [78, 152, 154, 155]). The outward Si-burning region compared to normal iron core-collapse supernovae produce higher [(Zn, Co, V)/Fe],[2] and smaller [(Mn, Cr)/Fe] ratios [155].

Core-collapse supernovae have long been regarded as astrophysical sites of r-process elements (e.g., [22, 60, 145, 162]). The r-process was thought to occur in the prompt explosion after the bounce. Recent numerical simulations, however, show that such an explosion does not occur. According to the recent simulations, neutrino-heating plays an important role in the explosion of core-collapse supernovae [67]. The neutrino-heating increases Y_e in supernovae. Wanajo [160] showed that the value of Y_e is typically larger than 0.4. The entropy is also not enough to synthesize r-process elements to the third peak. In models with proto-neutron star mass of 1.4 $M_\odot$, the entropy is 130 k_B nucleon^{-1} at 10 s after the bounce. They show that the mass of proto-neutron stars should be larger than 2.0 $M_\odot$, which is too massive compared to the typical mass of neutron stars. These results suggest that core-collapse supernovae are not a suitable site for the r-process. Supernovae from strongly magnetized and rapidly rotating progenitors currently are the only possibility to synthesize r-process elements to the third peak [100, 130, 168]. These types of supernovae would have low Y_e environment which is suitable for the r-process.

1.1.3.2 Neutron Star Mergers

Neutron star mergers can be other major sites for the enrichment of heavy elements (e.g., [149]). Observed binary pulsars are one of the indirect evidence of the existence of neutron star mergers. In the Milky Way, there are at least four binary pulsars that merge within the Hubble time [84]. Short gamma-ray bursts are also thought to be associated with neutron star mergers (e.g., [44]). By using these signatures, the rate of neutron star mergers is estimated to be 1000^{+9000}_{-990} Gpc^{-3}yr^{-1} [1].[3]

Advanced LIGO/Virgo detects the first event of a neutron star merger (GW170817) on August 17, 2017 [3]. They reported that two neutron stars with the total mass of 2.74 $M_\odot$ are merged at the distance of 40 Mpc from the Earth. The rate of neutron star mergers estimated from this observation is 1540^{+3200}_{-1220} Gpc^{-3}yr^{-1}. After the detection of gravitational waves, the counterpart of GW170817 is detected in multi-messenger (gamma-ray, X-ray, ultraviolet, optical, infrared, and radio) observations [2, 4, 5, 7,

---

[2][A/B] = $\log_{10}(N_A/N_B) - \log_{10}(N_A/N_B)_\odot$, where N_A and N_B are the number of the elements A and B, respectively.

[3]The units "Gpc" and "yr" mean gigaparsec (= 10^9 pc = 3.085×10^{27} cm) and year, respectively. One parsec (pc) is equal to 3.085×10^{18} cm. In this study, we also use "Mpc" and "kpc" which mean megaparsec (10^6 pc) and kiloparsec (10^3 pc), respectively.

8, 27, 31, 39, 54, 57, 73, 81, 82, 88, 94, 99, 111, 123, 135, 138, 148, 153, 156, 157]. Figure 1.4 shows timelines of the multi-messenger observations of GW170817. The short gamma-ray burst, GRB 170817A is observed after 1.74 s after the detection of the gravitational waves [2]. This observation confirms that neutron star mergers are progenitors of short gamma-ray bursts.

The optical counterpart, kilonova/macronova AT2017gfo/SSS17a is also observed after the detection of GW170817 (e.g., [135, 156]). At the early stages of the afterglow (<2 days), the blue optical component is dominated. After this phase, the color of AT2017gfo becomes rapidly redder. This light curve is different from those of supernovae. Tanaka et al. [147] performed radiative transfer simulations of the kilonova. They found that the ejecta of 0.03 $M_\odot$ which contain lanthanide elements power the emission. Central engine activities of neutron star mergers can also explain such observed light curves [89]. Future observations of the long-time evolution of afterglow from neutron star mergers can confirm that whether or not they produce r-process elements to the interstellar medium [164, 171].

Nucleosynthesis studies have shown that neutron star mergers can synthesize r-process elements with $A \geq 110$ (e.g., [149]). Dynamical ejecta from neutron star mergers produce materials with $Y_e < 0.1$. Such ejecta are suitable for synthesis of the heavy r-process elements. However, it is not enough to explain the universality of r-process elements between the solar r-process pattern and r-process rich metal-poor stars [136]. Newtonian simulations of neutron star mergers yield only heavier r-process elements ($A \gtrsim 110$). There is no observed evidence that there are stars which have only heavy r-process elements. Recent simulations with general relativity and neutrino transport showed shock heated ejecta have relatively higher Y_e ($0.1 < Y_e < 0.4$, [163]). Wanajo et al. [163] showed that neutron star mergers can synthesize r-process elements in the range $90 \lesssim A \lesssim 240$ (Fig. 1.5). This result suggests that neutron star mergers are the promising sources of r-process elements in the Universe.

1.2 The Local Group Galaxies

1.2.1 The Observed Local Group Galaxies

The Local Group consists of over 100 galaxies including the Milky Way [93]. These galaxies exist within 3 Mpc from the Sun. It is possible to resolve stellar populations of several galaxies in the Local Group because they are close to the Earth. The observations of chemodynamical properties of the Local Group galaxies help us understand the formation and evolution of the Local Group galaxies.

Star formation histories in the Local Group galaxies can be estimated from color-magnitude diagrams. Weisz et al. [165] showed that the Local Group dwarf galaxies have various star formation histories (Fig. 1.6). Their results suggest that average

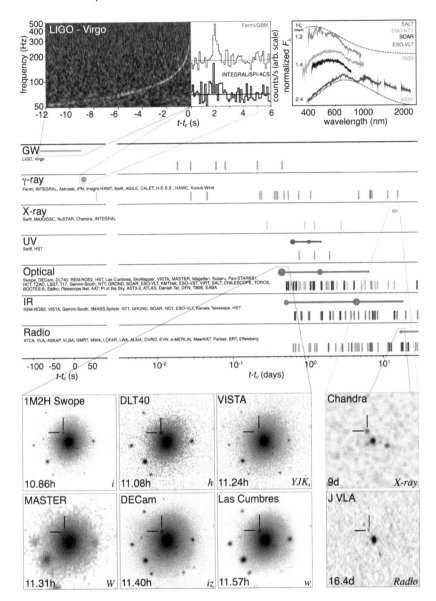

Fig. 1.4 The timeline of the multi-messenger observations of GW170817 [Fig. 2 of Abbott [4] reproduced under the terms of Creative Commons Attribution 3.0 license (https://creativecommons. org/licenses/by/3.0/)]. The observing facilities, instruments, and teams are listed at the beginning of the row. The solid circles denote representative observations. The vertical solid lines indicate the source was able to detect at least one instrument. The top left magnification inset shows that the frequency of gravitational waves detected by LIGO-Hanford and LIGO-Livingston as a function of time. The top middle inset represents the X-ray light curve. The top right inset denotes optical/near-infrared spectra observed by SALT (red curve, [20, 94]), ESO-NTT (green curve, [135]), the SOAR 4 m telescope (black curve, [99]), and ESO-VLT-XShooter (blue curve, [135]). Bottom panels illustrate optical, X-ray, and radio images of the counterpart of GW170817

Fig. 1.5 The comparison with solar r-process abundance patterns and yields of r-process elements from neutron star mergers (Fig. 4 of Wanajo et al. [163] reproduced by permission of the AAS). The top panel represents r-process abundance pattern in the ejecta of $Y_e = 0.09$ (the red curve), 0.14 (the light green curve), 0.19 (the magenta curve), 0.24 (the brown curve), 0.34 (the blue curve), and 0.44 (the purple curve). The bottom panel compares solar r-process abundances (black circles), and mass-averaged yields (the red curve)

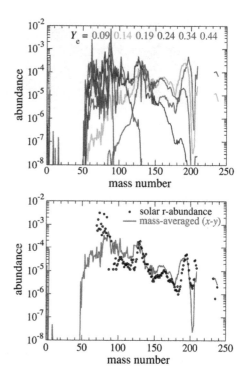

star formation histories in dwarf spheroidal galaxies exponentially decline with a timescale of ∼5 billion years. For dwarf irregulars, transition dwarfs, and dwarf ellipticals, star formation histories are characterized by the combination of an exponentially declining star formation history with timescales of 3–4 billion years and a constant star formation history after that. They also suggest that lower mass galaxies tend to stop the star formation in the early phase.

The detailed observations of the individual galaxy in the Local Group show that characteristic features in each galaxy. The intermediate age (1–10 billion years) stars dominate stellar populations in the Fornax dwarf spheroidal galaxy [32]. It also has a radial age gradient. Fractions of younger and more metal-rich populations are more abundant in its central region than the outer region. For the Sculptor dwarf spheroidal galaxy, old and metal-poor stars dominate stellar populations [33]. Its star formation history has a peak at 13–14 billion years ago and ends the star formation at 7 billion years ago. The Carina dwarf spheroidal galaxy has episodic star formation histories (e.g., [34]). The interaction with the Milky Way [112] or gas infall [80] would trigger the episodic star formation histories.

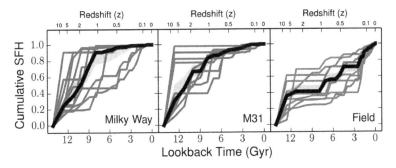

Fig. 1.6 Star formation histories of the Local Group dwarf galaxies (Fig. 11 of Weisz et al. [165] reproduced by permission of the AAS). From left to right, panels show the Milky Way satellites, M31 satellites, and field dwarf galaxies. The black solid curve shows the average star formation history of all galaxies in each field. The gray shaded region shows the error of the average star formation history. Colored curves show star formation histories of each galaxy

1.2.2 Chemical Abundances of Local Group Dwarf Galaxies

The chemical abundances of metal-poor stars in Local Group dwarf galaxies provide us with an imprint of evolutionary histories of galaxies. Metallicity distribution functions are one of the fundamental indicators for the chemical evolution of galaxies. Metallicity in Local Group dwarf galaxies has been measured by the empirical relation between the Ca II infrared triplet and [Fe/H], calibrated based on metallicities of globular clusters. Helmi et al. [58] found that metallicity distribution functions in Sculptor, Sextans, Fornax, and Carina dwarf spheroidal galaxies lack stars with [Fe/H] $\lesssim$ −3. Their results shed light on the question, where the extremely metal-poor stars (stars with [Fe/H] < −3) in the Milky Way halo were formed. On the other hand, Starkenburg et al. [139] showed that the widely-used liner empirical calibration of Ca II infrared triplet strongly deviates from their revised calibration at [Fe/H] < −2. Extremely metal-poor stars currently have been found in classical dwarf spheroidal galaxies with median and high-resolution spectroscopy (e.g., [46, 140, 146]). Besides, Kirby et al. [74] found 15 extremely metal-poor stars in seven ultra-faint dwarf galaxies. Norris et al. [108] reported that there is a star with [Fe/H] = −3.4 in the Boötes I ultra-faint dwarf galaxy. These observations of extremely metal-poor stars support the hierarchical structure formation scenario from Lambda (Λ) cold dark matter (CDM) cosmology. According to this scenario, the Milky Way halo is formed from clustering of smaller systems.

Average metallicity of galaxies correlates with their mass. The mass-metallicity relation of galaxies can be understood by the deepness of the gravitational potential well. More massive galaxies have deeper gravitational potential well and retain more gas and metals [37]. Dekel and Woo [38] predicted that metallicity (Z) and stellar mass (M_*) correlates with $Z \propto M_*^{0.4}$ by their analysis of the relation between supernova feedback and mass of galaxies. They found that this relation fits well with the observed mass-metallicity relation in Local Group dwarf galaxies.

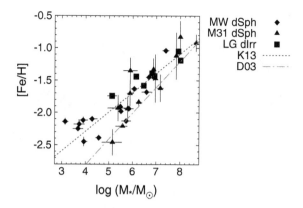

Fig. 1.7 The observed stellar mass-metallicity relation for the Milky Way dwarf spheroidal galaxies (diamonds), M31 dwarf spheroidal galaxies (triangles), and dwarf irregulars (squares) ([76] K13). The purple dashed and the sky-blue dot-dashed lines respectively denote the least square fitting of the samples of K13: $\langle[\text{Fe/H}]\rangle = (-1.69 \pm 0.04) + (0.30 \pm 0.02) \log\left(M_*/10^6 M_\odot\right)$ Dekel and Woo [38]: $[\text{Fe/H}] \propto M_*^{0.40}$ (D03)

Kirby et al. [76] measured metallicities of red-giants in 15 Milky Way dwarf spheroidal galaxies, seven Local Group dwarf irregulars, and 13 M31 dwarf spheroidal galaxies (Fig. 1.7). They found that the mass-metallicity relation roughly continues from galaxies with $M_* = 10^{3.5}$ to 10^{12} $M_\odot$. The derived relation using their sample ($M_* < 10^9$ $M_\odot$) is $Z \propto M_*^{0.30 \pm 0.02}$. They also found that both dwarf spheroidal galaxies and dwarf irregulars follow the same mass-metallicity relation. Although each galaxy has a different evolutionary history, all galaxies including Local Group galaxies lie in the same relation of mass and metallicity.

The elemental abundance ratios in galaxies give us clues to understand the astrophysical sites of elements and chemodynamical evolution of galaxies. Core-collapse supernovae synthesize the α-elements (O, Mg, Si, Ca, and Ti). O and Mg are synthesized during the hydrostatic He burning in massive stars. Si, Ca, and Ti are mostly produced during a core-collapse supernova explosion. On the other hand, both core-collapse supernovae and type Ia supernovae synthesize Fe. Type Ia supernovae produce Fe with a typical timescale of ~ 1 billion years (e.g., [87]). The onset of type Ia supernovae decreases the $[\alpha/\text{Fe}]$ ratio (e.g., [91, 150]). The point where $[\alpha/\text{Fe}]$ ratios start to decrease is called "knee." The $[\alpha/\text{Fe}]$ ratios indicate the speed of chemical evolution in galaxies. Figure 1.8 shows $[\alpha/\text{Fe}]$ as a function of $[\text{Fe/H}]$ in Local Group galaxies. At lower metallicity than the knee, $[\alpha/\text{Fe}]$ in dwarf galaxies follows that in the Milky Way halo. This result suggests that the enrichment of α-elements is similar in the Milky Way halo and Local Group dwarf galaxies at the early epoch of galaxy evolution. On the other hand, at higher metallicity, the metallicity of the knee in dwarf galaxies is lower than that of the Milky Way halo. Suda et al. [144] showed that positions of the knee are -2.1 (Fornax), -2.1 (Sculptor), -2.3 (Draco),

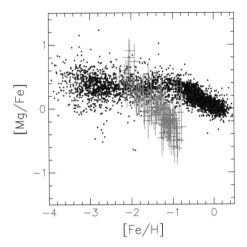

Fig. 1.8 Observed [Mg/Fe] ratios as a function of [Fe/H]. Red and black plots represent Mg abundances in the Sculptor dwarf spheroidal galaxy [75] and the Milky Way (SAGA database, [142–144, 172])

and −1.0 (Milky Way). This result suggests that the enrichment of Fe is slower than that of the Milky Way halo.

The abundances of the heaviest iron group element, zinc (Zn) show slightly different behavior from α-elements. Figure 1.9 represents [Zn/Fe] as functions of [Fe/H] in Local Group galaxies. At [Fe/H] $\lesssim$ −2.5, there is a decreasing trend toward higher metallicity (e.g., [25, 41, 101, 102, 121]). At higher metallicity, the decreasing trend continues in dwarf galaxies while there is a flat [Zn/Fe] ratio in the Milky Way halo. Skúuladóttir et al. [134] implied that there are star-to-star scatters of [Zn/Fe] ratios at [Fe/H] > −2.5. However, we have not yet well understood the enrichment of [Zn/Fe] ratios and astrophysical sites of Zn. We will discuss this issue in Chap. 4.

The abundance ratios of neutron-capture elements are very different from other elements. Figure 1.10 shows [Ba/Fe] as functions of [Fe/H]. Here we eliminate s-process contributions using the [Ba/Eu] ratio using the pure r-process [Ba/Eu] ratio: [Ba/Eu] = −0.89 [23]. In the Milky Way halo, the r-process component of [Ba/Fe] ratios show scatters over 3 dex in extremely metal-poor stars (e.g., [6, 13, 23, 45, 50, 59, 62, 64, 70, 95, 117, 122, 166]). This scatter indicates that r-process elements are produced at lower rates than the rate of whole supernovae when the spatial distribution of metallicity is not yet homogenized. On the other hand, dwarf spheroidal galaxies do not have stars with [Ba/Fe] > 1 at [Fe/H] $\lesssim$ −3 (e.g., [42, 47]). The r-process abundances in ultra-faint dwarf galaxies except for Reticulum II are also depleted compared to those of the Milky Way halo. Ji et al. [69] found that Reticulum II ultra-faint dwarf galaxy shows a strong enhancement of r-process elements. These observations help us understand the astrophysical sites of r-process elements and evolutionary histories of the Local Group galaxies (see Chaps. 5, 6, and 7).

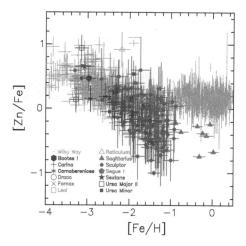

Fig. 1.9 The observed [Zn/Fe] ratios as functions of [Fe/H] [colored points, Fig. 1 of Hirai et al. [61] reproduced under the terms of Creative Commons Attribution 3.0 license (https://creativecommons. org/licenses/by/3.0/)]. The light gray points and the color points represent the Milky Way and the Local Group dwarf galaxies. We take the observed data for the Local Group dwarf galaxies as follows: Ursa Minor: Shetrone et al. [129], Sadakane et al. [120], Cohen and Huang [29], Ursa Major II: Frebel et al. [49], Sextans: Shetrone et al. [129], Honda et al. [63], Segue I: Frebel et al. [48], Sculptor: Shetrone et al. [128], Geisler et al. [51], Simon et al. [132], Skúladóttir et al. [133], Skúladóttir et al. [134], Sagittarius: Sbordone et al. [125], Reticulum II: Ji et al. [68], Fornax, Leo I: Shetrone et al. [128], Draco: Shetrone et al. [129], Cohen and Huang [28], Comaberenices: Frebel et al. [49], Carina: Shetrone et al. [128], Venn et al. [158], and Boötes I: Gilmore et al. [52]. Error bars shows the systematic and statistical errors given in each reference. All data are compiled using the SAGA database [142–144, 172]

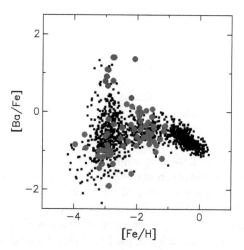

Fig. 1.10 The *r*-process component of [Ba/Fe] ratios as a function of [Fe/H] (SAGA database, [142–144, 172]). Ba abundances are corrected to be [Ba/Eu] = −0.89 at [Fe/H] > −2.75 to remove the *s*-process contribution. Black, blue, and red plots respectively represent [Ba/Fe] ratios in the Milky Way, Reticulum II, and the other Local Group dwarf galaxies (Carina, Draco, Leo I, Sculptor, Ursa Minor, Boötes I, Leo IV, Ursa Major II, Comaberenices, and Sextans)

1.3 Formation and Evolution of Galaxies

1.3.1 · Galaxy Formation

Early studies of galaxy formation were based on the observations of kinematics and colors (metallicities) of stars. Early observations showed that halo stars have fast velocity and low metallicity [109, 110, 118]. Eggen et al. [43] found that lower metallicity stars have higher eccentricities by the analysis of 221 stars. From their findings, they concluded that the galaxy is formed by the collapse of gas cloud within a few times 10^8 years. This monolithic collapse scenario had long been a standard scenario of galaxy formation.

On the other hand, Searle and Zinn [127] found that there is no radial dependence on metallicities of globular clusters in the outer halo. They also found that stars on the horizontal branch in these clusters show a large spread in the color distribution which indicates a broad range of age. These results suggested that the clustering of proto-galactic fragments forms the outer halo. Yoshii and Saio [173] found that metallicity and eccentricity were not correlated when they took into account the metal-poor stars with small proper motions. They showed that halo formation timescale is 3×10^9 years. Norris et al. [107] also found that ~20% of their objects are [Fe/H] ≤ -1.0 and the eccentricity less than 0.4. Chiba and Beers [26] concluded that there is no correlation between [Fe/H] and eccentricity. These results support the model of Searle and Zinn [127].

Hierarchical structure formation scenario based on ΛCDM cosmology successfully explains the properties of galaxies (e.g., [19]). White and Rees [167] showed that dark matter condensed around the time of recombination and clustered gravitationally to form larger systems. Development of tree method [14] for N-body calculation, smoothed particle hydrodynamics (SPH; [53, 85, 98]), and Eulerian adaptive mesh refinement (AMR; [17, 18]) enabled us to compute formation of galaxies in cosmological initial condition. Katz and Gunn [71] successfully made systems that resemble spiral galaxies by their N-body/hydrodynamic simulations. Steinmetz and Müller [141] computed the chemodynamical evolution of disk galaxies in a cosmological context. They showed that the disk of the galaxy has a metallicity gradient of $d(\log Z)/dr = -0.05$ kpc^{-1}.

Numerical simulations based on the hierarchical structure formation scenario currently become the dominant tool to study galaxy formation. Bekki and Chiba [15, 16] reproduced the observed properties of the Milky Way halo such as the metallicity-eccentricity relation, the radial density profile, and metallicity distribution functions by their chemodynamical simulations. Bullock and Johnston [21] found that satellites that contribute to the stellar halo accreted ~9 billion years ago in their hybrid semianalytic plus N-body model. This accretion time is older than that of surviving satellites (~5 billion years). Their result implied that stars in the inner stellar halo should have different chemical compositions from stars in the surviving satellite galaxies.

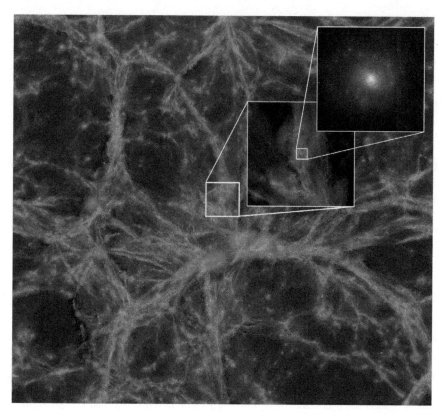

Fig. 1.11 A galaxy within a halo in the simulation (Fig. 1 of Schaye et al. [126] reproduced by permission of the Oxford University Press). From blue to red, color shows gas temperature. The inset denotes zoomed-in image of a galaxy formed in the simulation

Recent numerical simulations successfully reproduce several observed properties of galaxies. Figure 1.11 shows an example of such simulation (the EAGLE project [126]). Some of the simulations can resolve the scale of dwarf galaxies. Governato et al. [55] showed that the constant density profile in dwarf galaxies can be explained if strong outflows due to supernova feedback remove low-momentum gas. Ma et al. [86] reproduced the mass-metallicity relation of galaxies with $M_* = 10^4$–$10^{11}\ M_\odot$ in their simulations. Sawala et al. [124] performed cosmological hydrodynamic simulations of Local Group galaxies. Their simulated galaxies have the relation of stellar mass and velocity dispersion consistent with Local Group dwarf spheroidal galaxies.

1.3.2 Galactic Chemical Evolution

Galactic chemical evolution studies histories of the transformation from gas to stars and enrichment of elements in galaxies. Homogeneous one-zone models have long been used for the studies of galactic chemical evolution (e.g., [90, 151]). Essential ingredients of galactic chemical evolution models are stellar yields, lifetimes, initial mass function, star formation rates, and gas flows. One-zone chemical evolution models assume all metals are instantaneously mixed into the interstellar medium. These models derive average abundances of elements in a galaxy. One-zone models have been used as a powerful tool to study the astrophysical sites of elements (e.g., [78, 92, 115]).

In the early phases of galaxy evolution, chemical abundances in the interstellar medium were still inhomogeneous. Chemical abundances in very metal-poor stars reflect such inhomogeneities. Argast et al. [9, 10] introduced the stochastic chemical evolution model that can treat inhomogeneities of chemical abundances in the interstellar medium. In their model, the interstellar medium is divided into a cell of $(50 \text{ pc})^3$. Star formation takes place in probabilistically selected cells. To treat inhomogeneities, they assume that metals ejected from supernovae are mixed into the swept up materials.

Clarifying the enrichment of heavy elements can enhance our understanding of the astrophysical sites of elements and evolutionary history of galaxies. Understanding the enrichment of Zn will connect the chemical evolution from the high redshift to the present universe. Abundances of Zn have been used as a tracer of gas phase metallicity in damped Lyman-α systems at high redshift (e.g., [159, 169]). Although it is used as a tracer of metallicity, enrichment of Zn is not well understood. Kobayashi et al. [78] showed that hypernovae increase the average value of [Zn/Fe] ratios. However, they require hypernova rates ~ 10 times higher than those estimated from long gamma-ray bursts [56, 113]. Electron-capture supernovae may also contribute to the enrichment of Zn (see Chap. 4). For the first step to understand the enrichment of Zn, it is necessary to clarify the role of different kinds of supernovae.

For the r-process, neutron star mergers are thought to be a promising site of the r-process from nucleosynthetic studies (see Sect. 1.1.3.2). On the other hand, Argast et al. [11] failed to reproduce star-to-star scatters of r-process abundances in extremely metal-poor stars in the Milky Way halo by neutron star mergers due to their long merger times and low rates (high yields). Binary neutron stars typically need ~ 100 million years before they coalesce. Long merger times delay the production of r-process elements in a galaxy. Neutron star mergers typically eject $\sim 10^{-2}\ M_\odot$ of r-process elements with $\sim 10^3$ times lower rates of all core-collapse supernovae. This low rate causes extremely high abundance of r-process elements in high metallicity which are not seen in the observations. However, Argast et al. [11] did not consider the hierarchical structure formation (see Sect. 1.3.1) and metal mixing processes except for those in supernova remnants. Ishimaru et al. [65] showed that r-process elements increase at lower metallicity in less massive sub-halos in their one-zone chemical evolution model. Neutron star mergers occurred in halos with a size of dwarf spheroidal galaxies have r-process elements in [Fe/H] $\lesssim -2.5$. The central density not the total mass of halos affect the early star formation rates and r-process abundances (see Chap. 5).

The r-process elements and Zn could be used as a tracer of the efficiency of metal mixing in galaxies. Homogeneous chemical abundances in open star clusters are the direct evidence of metal mixing (e.g., [35, 36]). Roy and Kunth [119] studied the sites of metal mixing in gas-rich galaxies. They showed that drivers of metal mixing are divided into three different scales. On the largest scales (1–10 kpc), differential rotation homogenizes metals with a timescale of ≤ 1 billion years. On the intermediate scales (100–1000 pc), inter-cloud collisions play an important role in metal mixing. On the small scales (≤ 100 pc), turbulent diffusion mixes metals with a timescale of $\lesssim 1$ million years. However, which process is dominant in the context of galactic chemical evolution is not known. Comparison between heavy element abundances in simulations and metal-poor stars in Local Group galaxies can constrain the efficiency of metal mixing (see Chap. 6).

Galactic chemical evolution is related to the galaxy formation. We need to treat the chemical evolution in the context of the dynamical evolution of galaxies to understand evolutionary histories of galaxies fully. Galactic chemodynamical models can treat chemical evolution and galactic dynamics. Kawata and Gibson [72] developed a chemodynamical evolution code based on the N-body/SPH method. They showed that feedback from type Ia supernovae cannot be ignored in the evolution of elliptical galaxies. Revaz et al. [116] performed a series of chemodynamical simulations of isolated dwarf galaxy models. Their results suggested that the diversity of star formation histories and chemical abundances observed in Local Group galaxies are due to the intrinsic evolution of dwarf galaxies. Kobayashi and Nakasato [77] presented chemodynamical simulations of Milky Way-type galaxies. They showed that the enrichment histories of 13 kinds of elements and dynamical properties of stars that can be directly compared to the observations. Abundances of r-process elements could constrain the formation and evolution of Local Group galaxies. The r-process rich stars only exist in the Milky Way halo and Reticulum II ultra-faint dwarf galaxy. High resolution cosmological zoom-in simulations can directly answer this question (see Chap. 7).

1.4 This Book

This book concerns the enrichment of heavy elements (r-process elements and Zn) in dwarf galaxies. The first part of this book aims to clarify the contribution of neutron star mergers and different kinds of supernovae (core-collapse supernovae, hypernovae, electron-capture supernovae, and type Ia supernovae) to the enrichment of heavy elements (Chaps. 3–5). The second part of this book aims at constraining the formation and evolution of galaxies using heavy element abundances (Chaps. 6 and 7). This work covers enrichment of heavy elements in isolated dwarf galaxies

Fig. 1.12 Structure of this book

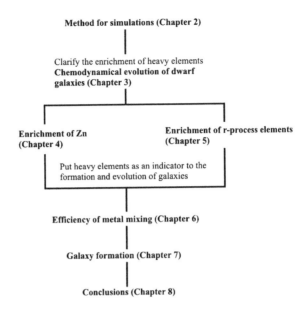

(Chaps. 3–6) and a cosmologically formed galaxy (Chap. 7). The key issues that will be discussed in this book are as follows.

- Chemodynamical evolution of dwarf galaxies (Chap. 3).
- Enrichment of Zn in dwarf galaxies (Chap. 4).
- Enrichment of r-process elements by neutron star mergers in dwarf spheroidal galaxies (Chap. 5).
- The relation of structures of halos, star formation histories, and the enrichment of r-process elements (Chap. 5).
- Efficiency of metal mixing in dwarf galaxies (Chap. 6).
- Connection between the enrichment of r-process elements and galaxy formation (Chap. 7).

Figure 1.12 illustrates the structure of this book. Chapter 2 describes methods for simulations of galaxies. Chapters 3–5, the first part of this book, study the enrichment of heavy elements in isolated dwarf galaxies. These works clarify the contributions of neutron star mergers and supernovae in the chemodynamical evolution of dwarf galaxies.

Chapter 3 This chapter introduces chemodynamical simulations of dwarf galaxies. Here we show radial profiles, star formation histories, metallicity distribution functions, mass-metallicity relations, and α-element abundances. We show that the isolated dwarf galaxy models can reproduce these quantities. We show that the central density and total mass of halos significantly affect the final properties of galaxies. Heating from supernovae efficiently reduce the star formation rates due to shallow gravitational potential well in dwarf galaxies. Metallicity distribution functions reflect the difference of star formation histories. We also show that ratios of α-elements to iron indicates the speed of chemical evolution with a timescale of ~ 1 billion years.

Chapter 4 In this chapter, we clarify the role of supernovae (core-collapse supernovae, electron-capture supernovae, hypernovae, and type Ia supernovae) to the enrichment of heavy elements. Zn is a nice tracer to identify the role of supernovae in the galactic chemical evolution. Here we newly put the effect of electron-capture supernovae into the simulations. We find that the ejecta from electron-capture supernovae can contribute to the formation of metal-poor stars with [Zn/Fe] $\gtrsim 0.5$. We also find that scatters of [Zn/Fe] seen in [Fe/H] $\gtrsim -2.5$ due to the contribution of Type Ia supernovae are the same as those of [Mg/Fe]. This chapter demonstrates that electron-capture supernovae or supernovae from low mass progenitors can be one of the contributors to the enrichment of Zn in galaxies.

Chapter 5 Neutron star mergers are one of the promising astrophysical site of the r-process. This chapter shows that neutron star mergers with a merger time of ~ 100 million years can contribute to the enrichment of r-process elements at [Fe/H] < -2.5 in a dwarf spheroidal galaxy. We find that this is because the metallicity does not increase due to the low star formation efficiency up to ~ 300 million years from the beginning of the star formation. We also find that galaxies with dynamical times with ~ 100 million years have star formation rates with $\lesssim 10^{-3}$ $M_\odot \mathrm{yr}^{-1}$ at 1 billion years from the beginning of the simulation. These galaxies produce stars with r-process elements in [Fe/H] $\lesssim -2.5$. On the other hand, r-process elements appear at higher metallicity in galaxies with dynamical times less than 100 million years. This chapter demonstrates that neutron star mergers can be a major astrophysical site of r-process elements in the context of the chemodynamical evolution of galaxies.

The second part of this book studies the efficiency of metal mixing and formation of galaxies on the point of enrichment of heavy elements. This part demonstrates abundances of heavy elements in metal-poor stars can be used as a tracer of the formation and evolution of galaxies.

Chapter 6 Chemical abundances of metal-poor stars help us understand the efficiency of metal mixing in galaxies. In this chapter, we newly implement the turbulence-induced metal mixing model to our chemodynamical simulations. We use Mg, Zn, and Ba as a tracer of the efficiency of metal mixing in dwarf galaxies. We find that models with scaling factor of metal diffusion less than 0.01 can produce abundances of heavy elements in metal-poor stars consistent with observations. This efficiency of metal mixing corresponds to the timescale of metal mixing less than 40 million years. This result implies that metals should be mixed with shorter timescales than the dynamical times of dwarf galaxies.

Chapter 7 The ratios of r-process elements to Fe in stars with [Fe/H] < -2.5 in the Local Group galaxies are characterized by (i) star-to-star scatters in the Milky Way halo, (ii) no r-process rich stars in dwarf spheroidal galaxies, (iii) enhanced r-process abundances in Reticulum II ultra-faint dwarf galaxy (Fig. 1.10). These difference may reflect the formation of Local Group galaxies. This chapter connects the enrichment of r-process elements and galaxy formation. Here we perform a series of high-resolution cosmological zoom-in simulations with metal mixing based on Chap. 6. In the cosmologically formed galaxy, there are star-to-star scatters of r-process abundances in [Fe/H] < -2. We find that r-process rich stars are formed in the halos with a gas mass of $\sim 10^6\ M_\odot$. In these small halos, ejecta from a neutron star merger enriches whole gas to high r-process abundances. This result suggests that the r-process enhanced stars observed in the Milky Way halo may come from halos with a size similar to the present ultra-faint dwarf galaxies.

This book is concluded in Chap. 8. Chapter 8 summarizes the scenario proposed in this book. We also show prospects of this work.

References

1. Abadie J et al (2010) TOPICAL REVIEW: predictions for the rates of compact binary coalescences observable by ground-based gravitational-wave detectors. Class Quantum Gravity 27(17):173001
2. Abbott BP et al (2017a) Gravitational waves and gamma-rays from a binary neutron star merger: GW170817 and GRB 170817A. Astrophys J Lett 848:L13
3. Abbott BP et al (2017b) GW170817: observation of gravitational waves from a binary neutron star inspiral. Phys Rev Lett 119(16):161101
4. Abbott BP et al (2017c) Multi-messenger observations of a binary neutron star merger. Astrophys J Lett 848:L12
5. Alexander KD, Berger E, Fong W, Williams PKG, Guidorzi C, Margutti R, Metzger BD, Annis J, Blanchard PK, Brout D, Brown DA, Chen H-Y, Chornock R, Cowperthwaite PS, Drout M, Eftekhari T, Frieman J, Holz DE, Nicholl M, Rest A, Sako M, Soares-Santos M, Villar VA (2017) The electromagnetic counterpart of the binary neutron star merger LIGO/Virgo GW170817. VI. Radio constraints on a relativistic jet and predictions for latetime emission from the Kilonova Ejecta. Astrophys J 848:L21
6. Aoki W, Suda T, Boyd RN, Kajino T, Famiano MA (2013) Explaining the Sr and Ba scatter in extremely metal-poor stars. Astrophys J Lett 766:L13
7. Arcavi I, Hosseinzadeh G, Howell DA, McCully C, Poznanski D, Kasen D, Barnes J, Zaltzman M, Vasylyev S, Maoz D, Valenti S (2017a) Optical emission from a kilonova following a gravitational-wave-detected neutron-star merger. Nature 551:64–66
8. Arcavi I, McCully C, Hosseinzadeh G, Howell DA, Vasylyev S, Poznanski D, Zaltzman M, Maoz D, Singer L, Valenti S, Kasen D, Barnes J, Piran T, Fong W-F (2017b) Optical follow-up of gravitational-wave events with Las Cumbres observatory. Astrophys J Lett 848:L33
9. Argast D, Samland M, Gerhard OE, Thielemann F-K (2000) Metal-poor halo stars as tracers of ISM mixing processes during halo formation. Astron Astrophys 356:873–887
10. Argast D, Samland M, Thielemann F-K, Gerhard OE (2002) Implications of O and Mg abundances in metal-poor halo stars for stellar iron yields. Astron Astrophys 388:842–860
11. Argast D, Samland M, Thielemann F-K, Qian Y-Z (2004) Neutron star mergers versus core-collapse supernovae as dominant r-process sites in the early Galaxy. Astron Astrophys 416:997–1011
12. Asplund M, Grevesse N, Sauval AJ, Scott P (2009) The chemical composition of the sun. Annu Rev Astron Astrophys 47:481–522
13. Barklem PS, Christlieb N, Beers TC, Hill V, Bessell MS, Holmberg J, Marsteller B, Rossi S, Zickgraf F-J, Reimers D (2005) The Hamburg/ESO R-process enhanced star survey (HERES). II. Spectroscopic analysis of the survey sample. Astron Astrophys 439:129–151

14. Barnes J, Hut P (1986) A hierarchical O(N log N) force-calculation algorithm. Nature 324:446–449

15. Bekki K, Chiba M (2000) Formation of the galactic stellar halo: origin of the metallicity-eccentricity relation. Astrophys J Lett 534:L89–L92

16. Bekki K, Chiba M (2001) Formation of the galactic stellar halo. I. Structure and kinematics. Astrophys J 558:666–686

17. Berger MJ, Colella P (1989) Local adaptive mesh refinement for shock hydrodynamics. J Comput Phys 82:64–84

18. Berger MJ, Oliger J (1984) Adaptive mesh refinement for hyperbolic partial differential equations. J Comput Phys 53:484–512

19. Blumenthal GR, Faber SM, Primack JR, Rees MJ (1984) Formation of galaxies and large-scale structure with cold dark matter. Nature 311:517–525

20. Buckley DAH, Andreoni I, Barway S, Cooke J, Crawford SM, Gorbovskoy E, Gromadski M, Lipunov V, Mao J, Potter SB, Pretorius ML, Pritchard TA, Romero-Colmenero E, Shara MM, Vaisanen P, Williams TB (2018) A comparison between SALT/SAAO observations and kilonova models for AT 2017gfo: the first electromagnetic counterpart of a gravitational wave transient-GW170817. Mon Not R Astron Soc Lett 474:L71–L75

21. Bullock JS, Johnston KV (2005) Tracing galaxy formation with stellar halos I. Methods. Astrophys J 635:931–949

22. Burbidge EM, Burbidge GR, Fowler WA, Hoyle F (1957) Synthesis of the elements in stars. Rev Mod Phys 29:547–650

23. Burris DL, Pilachowski CA, Armandroff TE, Sneden C, Cowan JJ, Roe H (2000) Neutron-capture elements in the early galaxy: insights from a large sample of metal-poor giants. Astrophys J 544:302–319

24. Cameron AGW (1957) Stellar evolution, nuclear astrophysics, and nucleogenesis. Chalk River report, CLR-41

25. Cayrel R, Depagne E, Spite M, Hill V, Spite F, François P, Plez B, Beers T, Primas F, Andersen J, Barbuy B, Bonifacio P, Molaro P, Nordström B (2004) First stars V-abundance patterns from C to Zn and supernova yields in the early Galaxy. Astron Astrophys 416:1117–1138

26. Chiba M, Beers TC (2000) Kinematics of metal-poor stars in the Galaxy. III. Formation of the stellar halo and thick disk as revealed from a large sample of nonkinematically selected stars. Astron J 119:2843–2865

27. Chornock R et al. (2017) The electromagnetic counterpart of the binary neutron star merger LIGO/Virgo GW170817. IV. Detection of near-infrared signatures of r-process nucleosynthesis with Gemini-South. Astrophys J Lett 848:L19

28. Cohen JG, Huang W (2009) The chemical evolution of the Draco dwarf spheroidal galaxy. Astrophys J 701:1053–1075

29. Cohen JG, Huang W (2010) The chemical evolution of the Ursa Minor dwarf spheroidal galaxy. Astrophys J 719:931–949

30. Cowan JJ, Sneden C, Burles S, Ivans II, Beers TC, Truran JW, Lawler JE, Primas F, Fuller GM, Pfeiffer B, Kratz K-L (2002) The chemical composition and age of the metal-poor halo star BD +17°3248. Astrophys J 572:861–879

31. Cowperthwaite PS et al (2017) The electromagnetic counterpart of the binary neutron star merger LIGO/Virgo GW170817. II. UV, optical, and near-infrared light curves and comparison to Kilonova models. Astrophys J Lett 848:L17

32. de Boer TJL, Tolstoy E, Hill V, Saha A, Olszewski EW, Mateo M, Starkenburg E, Battaglia G, Walker MG (2012a) The star formation and chemical evolution history of the Fornax dwarf spheroidal galaxy. Astron Astrophys 544:A73

33. de Boer TJL, Tolstoy E, Hill V, Saha A, Olsen K, Starkenburg E, Lemasle B, Irwin MJ, Battaglia G (2012b) The star formation and chemical evolution history of the Sculptor dwarf spheroidal galaxy. Astron Astrophys 539:A103

34. de Boer TJL, Tolstoy E, Lemasle B, Saha A, Olszewski EW, Mateo M, Irwin MJ, Battaglia G (2014) The episodic star formation history of the Carina dwarf spheroidal galaxy. Astron Astrophys 572:A10

35. De Silva GM, Freeman KC, Asplund M, Bland-Hawthorn J, Bessell MS, Collet R (2007a) Chemical homogeneity in collinder 261 and implications for chemical tagging. Astron J 133:1161–1175

36. De Silva GM, Freeman KC, Bland-Hawthorn J, Asplund M, Bessell MS (2007b) Chemically tagging the HR 1614 moving group. Astron J 133:694–704

37. Dekel A, Silk J (1986) The origin of dwarf galaxies, cold dark matter, and biased galaxy formation. Astron J 303:39–55

38. Dekel A, Woo J (2003) Feedback and the fundamental line of low-luminosity low-surface-brightness/dwarf galaxies. Mon Not R Astron Soc 344:1131–1144

39. Díaz MC et al (2017) Observations of the first electromagnetic counterpart to a gravitational-wave source by the TOROS collaboration. Astrophys J Lett 848:L29

40. Doherty CL, Gil-Pons P, Siess L, Lattanzio JC, Lau HHB (2015) Super- and massive AGB stars-IV. Final fates-initial-to-final mass relation. Mon Not R Astron Soc 446:2599–2612

41. Duffau S et al (2017) The Gaia-ESO survey: galactic evolution of sulphur and zinc. Astron Astrophys 604:A128

42. Duggan Gina E, Kirby Evan N, Andrievsky Serge M, Korotin Sergey A (2018) Neutron star mergers are the dominant source of the r-process in the early evolution of dwarf galaxies. Astrophys J 869(50):50

43. Eggen OJ, Lynden-Bell D, Sandage AR (1962) Evidence from the motions of old stars that the Galaxy collapsed. Astrophys J 136:748

44. Fernández R, Metzger BD (2016) Electromagnetic signatures of neutron star mergers in the advanced LIGO era. Annu Rev Nucl Part Sci 66:23–45

45. François P, Depagne E, Hill V, Spite M, Spite F, Plez B, Beers TC, Andersen J, James G, Barbuy B, Cayrel R, Bonifacio P, Molaro P, Nordström B, Primas F (2007) First stars. VIII. Enrichment of the neutron-capture elements in the early Galaxy. Astron Astrophys 476:935–950

46. Frebel A, Kirby EN, Simon JD (2010) Linking dwarf galaxies to halo building blocks with the most metal-poor star in Sculptor. Nature 464:72–75

47. Frebel A, Norris JE (2015) Near-field cosmology with extremely metal-poor stars. Annu Rev Astron Astrophys 53:631–688

48. Frebel A, Simon JD, Kirby EN (2014) Segue 1: an unevolved fossil galaxy from the early universe. Astrophys J 786:74

49. Frebel A, Simon JD, Geha M, Willman B (2010) High-resolution spectroscopy of extremely metal-poor stars in the least evolved galaxies: Ursa Major II and coma berenices. Astrophys J 708:560–583

50. Fulbright JP (2000) Abundances and kinematics of field halo and disk stars. I. Observational data and abundance analysis. Astrophys J 120:1841–1852

51. Geisler D, Smith VV, Wallerstein G, Gonzalez G, Charbonnel C (2005) "Sculptor-ing" the galaxy? The chemical compositions of red giants in the Sculptor dwarf spheroidal galaxy. Astrophys J 129:1428–1442

52. Gilmore G, Norris JE, Monaco L, Yong D, Wyse RFG, Geisler D (2013) Elemental abundances and their implications for the chemical enrichment of the Boötes I ultrafaint galaxy. Astrophys J 763:61

53. Gingold RA, Monaghan JJ (1977) Smoothed particle hydrodynamics-Theory and application to non-spherical stars. Mon Not R Astron Soc 181:375–389

54. Goldstein A, Veres P, Burns E, Briggs MS, Hamburg R, Kocevski D, Wilson-Hodge CA, Preece RD, Poolakkil S, Roberts OJ, Hui CM, Connaughton V, Racusin J, von Kienlin A, Dal Canton T, Christensen N, Littenberg T, Siellez K, Blackburn L, Broida J, Bissaldi E, Cleveland WH, Gibby MH, Giles MM, Kippen RM, McBreen S, McEnery J, Meegan CA, Paciesas WS, Stanbro M (2017) An ordinary short gamma-ray burst with extraordinary implications: fermi-GBM detection of GRB 170817A. Astrophys J Lett 848:L14

55. Governato F, Brook C, Mayer L, Brooks A, Rhee G, Wadsley J, Jonsson P, Willman B, Stinson G, Quinn T, Madau P (2010) Bulgeless dwarf galaxies and dark matter cores from supernova-driven outflows. Nature 463:203–206

56. Guetta D, Della Valle M (2007) On the rates of gamma-ray bursts and type Ib/c supernovae. Astrophys J Lett 657:L73–L76

57. Haggard D, Nynka M, Ruan JJ, Kalogera V, Cenko SB, Evans P, Kennea JA (2017) A deep Chandra X-Ray study of neutron star coalescence GW170817. Astrophys J Lett 848:L25

58. Helmi A, Irwin MJ, Tolstoy E, Battaglia G, Hill V, Jablonka P, Venn K, Shetrone M, Letarte B, Arimoto N, Abel T, François P, Kaufer A, Primas F, Sadakane K, Szeifert T (2006) A new view of the dwarf spheroidal satellites of the Milky Way from VLT FLAMES: where are the very metal-poor stars? Astrophys J Lett 651:L121–L124

59. Hill V, Plez B, Cayrel R, Beers TC, Nordström B, Andersen J, Spite M, Spite F, Barbuy B, Bonifacio P, Depagne E, François P, Primas F (2002) First stars. I. The extreme r-element rich, iron-poor halo giant CS 31082–001. Implications for the r-process site(s) and radioactive cosmochronology. Astron Astrophys 387:560–579

60. Hillebrandt W, Nomoto K, Wolff RG (1984) Supernova explosions of massive stars-The mass range 8 to 10 solar masses. Astron Astrophys 133:175–184

61. Hirai Y, Saitoh TR, Ishimaru Y, Wanajo S (2018) Enrichment of Zinc in galactic chemody-namical evolution models. Astrophys J 855(63):63

62. Honda S, Aoki W, Kajino T, Ando H, Beers TC, Izumiura H, Sadakane K, Takada-Hidai M (2004) Spectroscopic studies of extremely metal-poor stars with the Subaru High Dispersion Spectrograph. II. The r-process elements including thorium. Astrophys J 607:474–498

63. Honda S, Aoki W, Arimoto N, Sadakane K (2011) Enrichment of heavy elements in the red giant S 15–19 in the sextans dwarf spheroidal galaxy. Publ Astron Soc Jpn 63:523–529

64. Ishigaki MN, Aoki W, Chiba M (2013) Chemical abundances of the Milky Way thick disk and stellar halo. II. Sodium, iron-peak, and neutron-capture elements. Astrophys J 771:67

65. Ishimaru Y, Wanajo S, Prantzos N (2015) Neutron star mergers as the origin of r-process elements in the galactic halo based on the sub-halo clustering scenario. Astrophys J Lett 804:L35

66. Janka H-T, Melson T, Summa A (2016) Physics of core-collapse supernovae in three dimen-sions: a sneak preview. Annu Rev Nucl Part Sci 66:341–375

67. Janka H-T, Hanke F, Hüdepohl L, Marek A, Müller B, Obergaulinger M (2012) Core-collapse supernovae: reflections and directions. Prog Theor Exp Phys 1:01A309

68. Ji AP, Frebel A, Simon JD, Chiti A (2016a) Complete element abundances of nine stars in the r-process galaxy reticulum II. Astrophys J 830:93

69. Ji AP, Frebel A, Chiti A, Simon JD (2016b) R-process enrichment from a single event in an ancient dwarf galaxy. Nature 531:610–613

70. Johnson JA (2002) Abundances of 30 elements in 23 metal-poor stars. Astrophys J 139:219–247

71. Katz N, Gunn JE (1991) Dissipational galaxy formation. I-Effects of gasdynamics. Astrophys J 377:365–381

72. Kawata D, Gibson BK (2003) GCD+: a new chemodynamical approach to modelling super-novae and chemical enrichment in elliptical galaxies. Mon Not R Astron Soc 340:908–922

73. Kim S et al (2017) ALMA and GMRT constraints on the off-axis gamma-ray burst 170817A from the binary neutron star merger GW170817. Astrophys J Lett 850:L21

74. Kirby EN, Simon JD, Geha M, Guhathakurta P, Frebel A (2008) Uncovering extremely metal-poor stars in the Milky Way's ultrafaint dwarf spheroidal satellite galaxies. Astrophys J Lett 685(L43):L43

75. Kirby EN, Guhathakurta P, Simon JD, Geha MC, Rockosi CM, Sneden C, Cohen JG, Sohn ST, Majewski SR, Siegel M (2010) Multi-element abundance measurements from medium-resolution spectra. II. Catalog of stars in Milky Way dwarf satellite galaxies. Astrophys J Suppliment Ser 191:352–375

76. Kirby EN, Cohen JG, Guhathakurta P, Cheng L, Bullock JS, Gallazzi A (2013) The universal stellar mass-stellar metallicity relation for dwarf galaxies. Astrophys J 779:102

77. Kobayashi C, Nakasato N (2011) Chemodynamical simulations of the Milky Way galaxy. Astrophys J 729:16

78. Kobayashi C, Umeda H, Nomoto K, Tominaga N, Ohkubo T (2006) Galactic chemical evolution: carbon through zinc. Astrophys J 653:1145–1171

79. Kratz K-L, Bitouzet J-P, Thielemann F-K, Moeller P, Pfeiffer B (1993) Isotopic r-process abundances and nuclear structure far from stability-Implications for the r-process mechanism. Astrophys J 403:216–238

80. Lemasle B, Hill V, Tolstoy E, Venn KA, Shetrone MD, Irwin MJ, de Boer TJL, Starkenburg E, Salvadori S (2012) VLT/FLAMES spectroscopy of red giant branch stars in the Carina dwarf spheroidal galaxy. Astron Astrophys 538:A100

81. Li T et al (2018) Insight-HXMT observations of the first binary neutron star merger GW170817. Sci China Phys Mech Astron 61(3):31011

82. Lipunov VM et al (2017) MASTER optical detection of the first LIGO/Virgo neutron star binary merger GW170817. Astrophys J Lett 850:L1

83. Lodders K (2003) Solar system abundances and condensation temperatures of the elements. Astrophys J 591:1220–1247

84. Lorimer DR (2008) Binary and millisecond pulsars. Living Rev Relativ 11:8

85. Lucy LB (1977) A numerical approach to the testing of the fission hypothesis. Astron J 82:1013–1024

86. Ma X, Hopkins PF, Faucher-Giguère C-A, Zolman N, Muratov AL, Kereš D, Quataert E (2016) The origin and evolution of the galaxy mass-metallicity relation. Mon Not R Astron Soc 456:2140–2156

87. Maoz D, Mannucci F, Nelemans G (2014) Observational clues to the progenitors of type Ia supernovae. Annu Rev Astron Astrophys 52:107–170

88. Margutti R, Berger E, Fong W, Guidorzi C, Alexander KD, Metzger BD, Blanchard PK, Cowperthwaite PS, Chornock R, Eftekhari T, Nicholl M, Villar VA, Williams PKG, Annis J, Brown DA, Chen H, Doctor Z, Frieman JA, Holz DE, Sako M, Soares-Santos M (2017) The electromagnetic counterpart of the binary neutron star merger LIGO/Virgo GW170817. V. Rising X-Ray emission from an off-axis jet. Astrophys J 848:L20

89. Matsumoto T, Ioka K, Kisaka S, Nakar E (2018) Is the macronova in GW170817 powered by the central engine? Astrophys J 861(55):55

90. Matteucci F (ed) (2001) The chemical evolution of the Galaxy, vol 253. Astrophysics and Space Science Library

91. Matteucci F, Brocato E (1990) Metallicity distribution and abundance ratios in the stars of the Galactic bulge. Astrophys J 365:539–543

92. Matteucci F, Romano D, Arcones A, Korobkin O, Rosswog S (2014) Europium production: neutron star mergers versus core-collapse supernovae. Mon Not R Astron Soc 438:2177–2185

93. McConnachie AW (2012) The observed properties of dwarf galaxies in and around the local group. Astron J 144:4

94. McCully C, Hiramatsu D, Howell DA, Hosseinzadeh G, Arcavi I, Kasen D, Barnes J, Shara MM, Williams TB, Väisänen P, Potter SB, Romero- Colmenero E, Crawford SM, Buckley DAH, Cooke J, Andreoni I, Pritchard TA, Mao J, Gromadzki M, Burke J (2017) The rapid reddening and featureless optical spectra of the optical counterpart of GW170817, AT 2017gfo, during the first four days. Astrophys J Lett 848:L32

95. McWilliam A, Preston GW, Sneden C, Searle L (1995) Spectroscopic analysis of 33 of the most metal poor stars. II. Astron J 109:2757

96. Miyaji S, Nomoto K (1987) On the collapse of 8–10 solar mass stars due to electron capture. Astron J 318:307–315

97. Miyaji S, Nomoto K, Yokoi K, Sugimoto D (1980) Supernova triggered by electron captures. Publ Astron Soc Jpn 32:303

98. Monaghan JJ, Lattanzio JC (1985) A refined particle method for astrophysical problems. Astron Astrophys 149:135–143

99. Nicholl M et al (2017) The electromagnetic counterpart of the binary neutron star merger LIGO/Virgo GW170817. III. Optical and UV spectra of a blue kilonova from fast polar ejecta. Astrophys J Lett 848:L18

100. Nishimura N, Takiwaki T, Thielemann F-K (2015) The r-process nucleosynthesis in the various jet-like explosions of magnetorotational core-collapse supernovae. Astrophys J 810:109

101. Nissen PE, Chen YQ, Asplund M, Pettini M (2004) Sulphur and zinc abundances in Galactic stars and damped Lya systems. Astron Astrophys 415:993–1007

102. Nissen PE, Akerman C, Asplund M, Fabbian D, Kerber F, Kaufl HU, Pettini M (2007) Sulphur and zinc abundances in Galactic halo stars revisited. Astron Astrophys 469:319–330

103. Nomoto K (1984) Evolution of 8–10 solar mass stars toward electron capture supernovae. I-Formation of electron-degenerate O + Ne + Mg cores. Astrophys J 277:791–805

104. Nomoto K (1987) Evolution of 8–10 solar mass stars toward electron capture supernovae. II-Collapse of an O + Ne + Mg core. Astrophys J 322:206–214

105. Nomoto K, Kobayashi C, Tominaga N (2013) Nucleosynthesis in stars and the chemical enrichment of galaxies. Annu Rev Astron Astrophys 51:457–509

106. Nomoto K, Sugimoto D, Sparks WM, Fesen RA, Gull TR, Miyaji S (1982) The Crab Nebula's progenitor. Nature 299:803–805

107. Norris J, Bessell MS, Pickles AJ (1985) Population studies. I-The Bidelman-MacConnell 'weak-metal' stars. Astrophys J Suppl Ser 58:463–492

108. Norris JE, Gilmore G, Wyse RFG, Wilkinson MI, Belokurov V, Evans NW, Zucker DB (2008) The abundance spread in the Boötes I dwarf spheroidal galaxy. Astrophys J Lett 689:L113

109. Oort JH (1922) Some peculiarities in the motion of stars of high velocity. Bull Astron Inst Neth 1:133

110. Oort JH (1926) The stars of high velocity. PhD thesis. Publications of the Kapteyn Astronomical Laboratory Groningen, vol 40, pp 1–75

111. Pian E et al (2017) Spectroscopic identification of r-process nucleosynthesis in a double neutron-star merger. Nature 551:67–70

112. Piatek S, Pryor C, Olszewski EW, Harris HC, Mateo M, Minniti D, Tinney CG (2003) Proper motions of dwarf spheroidal galaxies from hubble space telescope imaging. II. Measurement for carina. Astron J 126:2346–2361

113. Podsiadlowski P, Langer N, Poelarends AJT, Rappaport S, Heger A, Pfahl E (2004a) The effects of binary evolution on the dynamics of core collapse and neutron star kicks. Astron J 612:1044–1051

114. Podsiadlowski P, Mazzali PA, Nomoto K, Lazzati D, Cappellaro E (2004b) The rates of hypernovae and gamma-ray bursts: implications for their progenitors. Astrophys J Lett 607:L17–L20

115. Prantzos N, Abia C, Limongi M, Chieffi A, Cristallo S (2018) Chemical evolution with rotating massive star yields-I. The solar neighbourhood and the s-process elements. Mon Not R Astron Soc 476:3432–3459

116. Revaz Y, Jablonka P, Sawala T, Hill V, Letarte B, Irwin M, Battaglia G, Helmi A, Shetrone MD, Tolstoy E, Venn KA (2009) The dynamical and chemical evolution of dwarf spheroidal galaxies. Astron Astrophys 501:189–206

117. Roederer IU, Preston GW, Thompson IB, Shectman SA, Sneden C, Burley GS, Kelson DD (2014) A search for stars of very low metal abundance. VI. Detailed abundances of 313 metal-poor stars. Astron J 147:136

118. Roman NG (1955) A catalogue of high-velocity stars. Astrophys J Suppl 2:195

119. Roy J-R, Kunth D (1995) Dispersal and mixing of oxygen in the interstellar medium of gas-rich galaxies. Astron Astrophys 294:432–442

120. Sadakane K, Arimoto N, Ikuta C, Aoki W, Jablonka P, Tajitsu A (2004) Subaru/HDS abundances in three giant stars in the Ursa Minor dwarf spheroidal galaxy. Publ Astron Soc Jpn 56:1041–1058

121. Saito Y-J, Takada-Hidai M, Honda S, Takeda Y (2009) Chemical evolution of zinc in the galaxy. Publ Astron Soc Jpn 61:549–561

122. Sakari CM, Placco VM, Farrell EM, Roederer IU, Wallerstein G, Beers TC, Ezzeddine R, Frebel A, Hansen T, Holmbeck EM, Sneden C, Cowan JJ, Venn KA, Christopher ED, Matijevič G, Wyse RFG, Bland-Hawthorn J, Chiappini C, Freeman KC, Gibson BK, Grebel EK, Helmi A, Kordopatis G, Kunder A, Navarro J, Reid W, Seabroke G, Steinmetz M, Watson F (2018)

The R-process alliance: first release from the northern search for r-process-enhanced metalpoor stars in the galactic halo. Astrophys J 868(110):110

123. Savchenko V, Ferrigno C, Kuulkers E, Bazzano A, Bozzo E, Brandt S, Chenevez J, Courvoisier TJ-L, Diehl R, Domingo A, Hanlon L, Jourdain E, von Kienlin A, Laurent P, Lebrun F, Lutovinov A, Martin-Carrillo A, Mereghetti S, Natalucci L, Rodi J, Roques J-P, Sunyaev R, Ubertini P (2017) INTEGRAL detection of the first prompt gamma-ray signal coincident with the gravitational-wave event GW170817. Astrophys J Lett 848:L15

124. Sawala T, Frenk CS, Fattahi A, Navarro JF, Bower RG, Crain RA, Dalla Vecchia C, Furlong M, Helly JC, Jenkins A, Oman KA, Schaller M, Schaye J, Theuns T, Trayford J, White SDM (2016) The APOSTLE simulations: solutions to the local group's cosmic puzzles. Mon Not R Astron Soc 457:1931–1943

125. Sbordone L, Bonifacio P, Buonanno R, Marconi G, Monaco L, Zaggia S (2007) The exotic chemical composition of the Sagittarius dwarf spheroidal galaxy. Astron Astrophys 465:815–824

126. Schaye J, Crain RA, Bower RG, Furlong M, Schaller M, Theuns T, Dalla Vecchia C, Frenk CS, McCarthy IG, Helly JC, Jenkins A, Rosas- Guevara YM, White SDM, Baes M, Booth CM, Camps P, Navarro JF, Qu Y, Rahmati A, Sawala T, Thomas PA, Trayford J (2015) The EAGLE project: simulating the evolution and assembly of galaxies and their environments. Mon Not R Astron Soc 446:521–554

127. Searle L, Zinn R (1978) Compositions of halo clusters and the formation of the galactic halo. Astrophys J 225:357–379

128. Shetrone M, Venn KA, Tolstoy E, Primas F, Hill V, Kaufer A (2003) VLT/UVES abundances in four nearby dwarf spheroidal galaxies. I. Nucleosynthesis and abundance ratios. Astrophys J 125:684–706

129. Shetrone MD, Côté P, Sargent WLW (2001) Abundance patterns in the Draco, Sextans, and Ursa Minor Dwarf spheroidal galaxies. Astrophys J 548:592–608

130. Shibagaki S, Kajino T, Mathews GJ, Chiba S, Nishimura S, Lorusso G (2016) Relative contributions of the weak, main, and fission-recycling r-process. Astrophys J 816:79

131. Simmerer J, Sneden C, Cowan JJ, Collier J, Woolf VM, Lawler JE (2004) The rise of the s-process in the galaxy. Astrophys J 617:1091–1114

132. Simon JD, Jacobson HR, Frebel A, Thompson IB, Adams JJ, Shectman SA (2015) Chemical signatures of the first supernovae in the Sculptor dwarf spheroidal galaxy. Astrophys J 802:93

133. Skúladóttir Á, Tolstoy E, Salvadori S, Hill V, Pettini M, Shetrone MD, Starkenburg E (2015) The first carbon-enhanced metal-poor star found in the Sculptor dwarf spheroidal. Astron Astrophys 574:A129

134. Skúladóttir Á, Tolstoy E, Salvadori S, Hill V, Pettini M (2017) Zinc abundances in the Sculptor dwarf spheroidal galaxy. Astron Astrophys 606:A71

135. Smartt SJ et al (2017) A kilonova as the electromagnetic counterpart to a gravitational-wave source. Nature 551:75–79

136. Sneden C, Cowan JJ, Gallino R (2008) Neutron-capture elements in the early galaxy. Annu Rev Astron Astrophys 46:241–288

137. Sneden C, Cowan JJ, Lawler JE, Ivans II, Burles S, Beers TC, Primas F, Hill V, Truran JW, Fuller GM, Pfeiffer B, Kratz K-L (2003) The extremely metal-poor, neutron capture-rich star CS 22892–052: a comprehensive abundance analysis. Astrophys J 591:936–953

138. Soares-Santos M et al (2017) The electromagnetic counterpart of the binary neutron star merger LIGO/Virgo GW170817. I. Discovery of the optical counterpart using the dark energy camera. Astrophys J Lett 848:L16

139. Starkenburg E, Hill V, Tolstoy E, González Hernández JI, Irwin M, Helmi A, Battaglia G, Jablonka P, Tafelmeyer M, Shetrone M, Venn K, de Boer T (2010) The NIR Ca ii triplet at low metallicity. Searching for extremely lowmetallicity stars in classical dwarf galaxies. Astron Astrophys 513:A34

140. Starkenburg E, Hill V, Tolstoy E, François P, Irwin MJ, Boschman L, Venn KA, de Boer TJL, Lemasle B, Jablonka P, Battaglia G, Groot P, Kaper L (2013) The extremely low-metallicity tail of the Sculptor dwarf spheroidal galaxy. Astron Astrophys 549:A88

141. Steinmetz M, Müller E (1994) The formation of disk galaxies in a cosmological context: populations, metallicities and metallicity gradients. Astron Astrophys 281:L97–L100
142. Suda T, Katsuta Y, Yamada S, Suwa T, Ishizuka C, Komiya Y, Sorai K, Aikawa M, Fujimoto MY (2008) Stellar abundances for the galactic archeology (SAGA) database-compilation of the characteristics of known extremely metal-poor stars. Publ Astron Soc Jpn 60:1159–1171
143. Suda T, Yamada S, Katsuta Y, Komiya Y, Ishizuka C, Aoki W, Fujimoto MY (2011) The stellar abundances for galactic archaeology (SAGA) data base-II. Implications for mixing and nucleosynthesis in extremely metal-poor stars and chemical enrichment of the Galaxy. Mon Not R Astron Soc 412:843–874
144. Suda T, Hidaka J, Aoki W, Katsuta Y, Yamada S, Fujimoto MY, Ohtani Y, Masuyama M, Noda K, Wada K (2017) Stellar abundances for galactic archaeology database. IV. Compilation of stars in dwarf galaxies. Publ Astron Soc Jpn 69:76
145. Sumiyoshi K, Terasawa M, Mathews GJ, Kajino T, Yamada S, Suzuki H (2001) r-process in prompt supernova explosions revisited. Strophysical J 562:880–886
146. Tafelmeyer M, Jablonka P, Hill V, Shetrone M, Tolstoy E, Irwin MJ, Battaglia G, Helmi A, Starkenburg E, Venn KA, Abel T, François P, Kaufer A, North P, Primas F, Szeifert T (2010) Extremely metal-poor stars in classical dwarf spheroidal galaxies: Fornax, Sculptor, and Sextans. Astron Astrophys 524:A58
147. Tanaka M et al (2017) Kilonova from post-merger ejecta as an optical and near-Infrared counterpart of GW170817. Publ Astron Soc Jpn 69:102
148. Tanvir NR et al (2017) The emergence of a lanthanide-rich kilonova following the merger of two neutron stars. Astrophys J Lett 848:L27
149. Thielemann F-K, Eichler M, Panov IV, Wehmeyer B (2017) Neutron star mergers and nucleosynthesis of heavy elements. Annu Rev Nucl Part Sci 67:253–274
150. Tinsley BM (1979) Stellar lifetimes and abundance ratios in chemical evolution. Astrophys J 229:1046–1056
151. Tinsley BM (1980) Evolution of the stars and gas in galaxies. Fundam Cosm Phys 5:287–388
152. Tominaga N, Umeda H, Nomoto K (2007) Supernova nucleosynthesis in population III 13–50 M_{solar} stars and abundance patterns of extremely metalpoor stars. Astrophys J 660:516–540
153. Troja E et al (2017) The X-ray counterpart to the gravitational-wave event GW170817. Nature 551:71–74
154. Umeda H, Nomoto K (2002) Nucleosynthesis of zinc and iron peak elements in population III type II supernovae: comparison with abundances of very metal poor halo stars. Astrophys J 565:385–404
155. Umeda H, Nomoto K (2005) Variations in the abundance pattern of extremely metal-poor stars and nucleosynthesis in population III supernovae. Astrophys J 619:427–445
156. Utsumi Y et al (2017) J-GEM observations of an electromagnetic counterpart to the neutron star merger GW170817. Publ Astron Soc Jpn 69:101
157. Valenti S, David J, Sand S, Yang E, Cappellaro L, Tartaglia A, Corsi SW, Jha DE, Haislip Reichart J, Kouprianov V (2017) The discovery of the electromagnetic counterpart of GW170817: kilonova AT 2017gfo/DLT17ck. Astrophys J Lett 848:L24
158. Venn KA, Shetrone MD, Irwin MJ, Hill V, Jablonka P, Tolstoy E, Lemasle B, Divell M, Starkenburg E, Letarte B, Baldner C, Battaglia G, Helmi A, Kaufer A, Primas F (2012) Nucleosynthesis and the inhomogeneous chemical evolution of the carina dwarf galaxy. Astrophys J 751:102
159. Vladilo G, Abate C, Yin J, Cescutti G, Matteucci F (2011) Silicon depletion in damped Ly a systems. The S/Zn method. Astron Astrophys 530:A33
160. Wanajo S (2013) The r-process in Proto-neutron-star Wind Revisited. Astrophys J Lett 770:L22
161. Wanajo S, Janka H-T, Müller B (2011) Electron-capture supernovae as the origin of elements beyond iron. Astrophys J Lett 726:L15
162. Wanajo S, Tamamura M, Itoh N, Nomoto K, Ishimaru Y, Beers TC, Nozawa S (2003) The r-process in supernova explosions from the collapse of O-Ne-Mg cores. Astrophys J 593:968–979

163. Wanajo S, Sekiguchi Y, Nishimura N, Kiuchi K, Kyutoku K, Shibata M (2014) Production of all the r-process nuclides in the dynamical ejecta of neutron star mergers. Astrophys J Lett 789:L39

164. Wanajo S (2018) Physical conditions for the r-process. I. Radioactive energy sources of kilonovae. Astrophys J 868:65

165. Weisz DR, Dolphin AE, Skillman ED, Holtzman J, Gilbert KM, Dalcanton JJ, Williams BF (2014) The star formation histories of local group dwarf galaxies. I. Hubble space telescope/wide field planetary camera 2 observations. Astrophys J 789:147

166. Westin J, Sneden C, Gustafsson B, Cowan JJ (2000) The r-process-enriched low-metallicity giant HD 115444. Astrophys J 530:783–799

167. White SDM, Rees MJ (1978) Core condensation in heavy halos-A twostage theory for galaxy formation and clustering. Mon Not R Astron Soc 183:341–358

168. Winteler C, Käppeli R, Perego A, Arcones A, Vasset N, Nishimura N, Liebendörfer M, Thielemann F-K (2012) Magnetorotationally driven supernovae as the origin of early galaxy r-process elements? Astrophys J Lett 750:L22

169. Wolfe AM, Gawiser E, Prochaska JX (2005) Damped Ly a systems. Annu Rev Astron Astrophys 43:861–918

170. Woosley SE, Weaver TA (1995) The evolution and explosion of massive stars. II. Explosive hydrodynamics and nucleosynthesis. Astrophys J 101:181

171. Wu M-R, Barnes J, Martinez-Pinedo G, Metzger BD (2019) Fingerprints of heavy element nucleosynthesis in the late-time lightcurves of kilonovae. Phys Rev Lett 122(6):062701

172. Yamada S, Suda T, Komiya Y, Aoki W, Fujimoto MY (2013) The stellar abundances for galactic archaeology (SAGA) database-III. Analysis of enrichment histories for elements and two modes of star formation during the early evolution of the MilkyWay. Mon Not R Astron Soc 436:1362–1380

173. Yoshii Y, Saio H (1979) Kinematics of the old stars and initial contraction of the galaxy. Publ Astron Soc Jpn 31:339–368

Chapter 2
Methods for Simulations of Galaxy Formation

Abstract The N-body/hydrodynamic simulations are powerful tools to conduct the study of the formation and evolution of galaxies. Several different groups have been developed codes for simulating galaxy formation. Here we describe the N-body and smoothed particle hydrodynamics (SPH) code, ASURA. Simulations of galaxy formation cannot resolve each star. We thus implement star formation and supernova feedback recipes in a simple stellar population approximation. We put the effects of iron-core collapse supernovae, hypernovae, electron-capture supernovae, neutron star mergers, type Ia supernovae, and asymptotic giant branch stars into the code. In this chapter, we also describe metal mixing models for SPH simulations [Contents in this chapter have been in part published in Hirai et al. (Astrophys J 814:41, 2015 [30]), (Astrophys J 855:63, 2018 [31]) reproduced by permission of the AAS].

2.1 The N-body and Smoothed Particle Hydrodynamics Code, ASURA

The N-body/hydrodynamic simulations have been mainly used to study the formation and evolution of galaxies. Various codes for galaxy formation with a different scheme for hydrodynamics have been developed, e.g., for smoothed particle hydrodynamics (SPH): GADGET [89, 91], GASOLINE [103, 104], and ASURA [79, 80], for adaptive mesh refinement: ART [40, 41, 72], ENZO [5], and RAMSES [98], and for moving mesh: AREPO [90].

Here we describe an N-body/SPH code, ASURA adopted in models described in this book. The ASURA code adopts dark matter, star, and gas particles. Dark matter and star particles are treated as collisionless particles. On the other hand, gas particles are treated as SPH particles. Dark matter particles are used to compute gravity. Star particles produce energy and heavy elements from supernovae and neutron star mergers. Gas particles are used to compute hydrodynamics.

The ASURA code solves hydrodynamics using SPH (e.g., [24, 51, 56, 57]). In SPH, the equation of motion and the energy equation of ith particle are described as

© Springer Nature Singapore Pte Ltd. 2019
Y. Hirai, *Understanding the Enrichment of Heavy Elements by the Chemodynamical Evolution Models of Dwarf Galaxies*, Springer Theses, https://doi.org/10.1007/978-981-13-7884-3_2

$$\frac{d^2 r_i}{dt^2} = -\sum_j m_j \left(\frac{P_i}{\rho_i^2} + \frac{P_j}{\rho_j^2} \right) \nabla W_{ij}(h), \tag{2.1}$$

$$\frac{du_i}{dt} = \sum_j m_j \frac{P_i}{\rho_i^2} v_{ij} \cdot \nabla W_{ij}(h), \tag{2.2}$$

where r_i, m_i, P_i, ρ_i, $v_{ij} = v_i - v_j$, $W_{ij} = W(|r_i - r_j|, h)$, h, and u_i are the position, mass, pressure, density, velocity, kernel function, smoothing length, and internal energy of gas particles. The physical quantity, f_i is described as follows,

$$f_i = \sum_j m_j \frac{f_j}{\rho_j} W_{ij}(h). \tag{2.3}$$

The density is thus described in the following way,

$$\rho_i = \sum_j m_j W_{ij}(h). \tag{2.4}$$

For SPH, a widely used kernel function is a cubic spline function:

$$W(r, h) = \frac{3}{2\pi h^3} \begin{cases} \frac{2}{3} - \left(\frac{r}{h}\right)^2 + \frac{1}{2}\left(\frac{r}{h}\right)^3 & \left(0 \le \frac{r}{h} < 1\right), \\ \frac{1}{6}\left[2 - \left(\frac{r}{h}\right)\right]^3 & \left(1 \le \frac{r}{h} < 2\right), \\ 0 & \left(\frac{r}{h} > 2\right). \end{cases} \tag{2.5}$$

These equations close by adopting the equation of state,

$$P = (\gamma - 1)\rho u, \tag{2.6}$$

where γ represents the specific heat ratio. SPH requires continuity and differentiability of the mass density. These requirements break down at a contact discontinuity. The formulation of SPH, therefore, cannot correctly treat a region with a contact discontinuity. To properly treat it, several authors have developed modified formulations of SPH such as the density-independent formulation of SPH (DISPH, [77]), the pressure-entropy SPH [33], and SPH with pseudo-density [111]. DISPH adopts the internal energy density for volume elements while the standard SPH adopts the mass density. The equations of motion and energy in DISPH are described as

$$m_i \frac{dv_i}{dt} = -(\gamma - 1) \sum_j U_i U_j \left(\frac{1}{q_i} + \frac{1}{q_j} \right) \nabla \tilde{W}_{ij}, \tag{2.7}$$

$$\frac{dU_i}{dt} = (\gamma - 1) \sum_j \frac{U_i U_j}{q_i} \boldsymbol{v}_{ij} \cdot \nabla \tilde{W}_{ij}, \tag{2.8}$$

where $q_j u_j$, $U_j = m_j u_j$, and $\nabla \tilde{W}_{ij} = 0.5[\nabla W_{ij}(h_i) + \nabla W_{ij}(h_j)]$. Saitoh and Makino [77] pointed out that DISPH can give more accurate results for tests with a contact discontinuity such as Kelvin-Helmholtz and Rayleigh-Taylor instabilities, the shock tube, and point-like explosion. Saitoh and Makino [78] showed that the DISPH produces the entropy core in simulations of galaxy clusters while the standard SPH cannot produce such cores. These cores are also produced in moving mesh and mesh free schemes. On the other hand, Hopkins [33] showed that the DISPH and the standard SPH produce similar time-averaged star formation rates in isolated galaxies. We choose widely used standard SPH method in Chaps. 3 and 5. We adopt DISPH in Chaps. 4, 6, and 7.

The ASURA code computes gravity using a tree method [3] to decrease the computation cost. The tree method hierarchically divides space into cubes. Following the opening angle criterion, forces from closer particles are directly computed while forces from a larger distance are computed with particles grouped. We set the value of 0.5 for the opening angles in our simulations. The ASURA code adopts the Phantom-GRAPE library [96, 97] to accelerate computations of gravity. This library uses a Single Instruction, Multiple Data (SIMD) instruction set, Advanced Vector Extensions (AVX) to accelerate force calculations. The code is parallelized following Makino [53].

The second-order scheme is adopted for time integration [78] with the individual-block time step [52, 55]. The ASURA code uses the algorithm of fully asynchronous split time-integrator (FAST) to decrease the cost of computation for integration of a fluid system of self-gravity [76]. The code adopts the time-step limiter [75]. The timesteps of a particle are limited to be at most four times longer than those of neighbor particles. This limiter enables us to correctly compute the evolution of regions with strong shock. It uses a cooling and heating function from $10-10^9$ K dependent on the metallicity [19, 20]. Ultra-violet background heating is implemented using a table presented in Haardt and Madau [27]. The code adopts the hydrogen self-shielding models following the fitting function shown in Rahmati et al. [68].

Star formation models in ASURA require three conditions for gas particles to be star particles: (1) gas particles are converging ($\nabla \cdot \mathbf{v} < 0$), (2) the density exceeds the threshold density (n_{th}), and (3) the temperature is below the threshold temperature (T_{th}). Gas particles satisfied these conditions become a candidate to form star particles. The ASURA then generates random numbers and compares with the value,

$$p = \frac{m_{gas}}{m_\star} \left\{ 1 - \exp\left(-c_\star \frac{dt}{t_{dyn}} \right) \right\}, \tag{2.9}$$

Table 2.1 Parameters adopted in the simulations

Quantity	Symbols	Fiducial value	Variation
Dimensionless star formation efficiency parameter	$c_\star$	0.033	0.033, 0.5
Threshold density of star formation	n_{th}	$100\ cm^{-3}$	$0.1–100\ cm^{-3}$
Threshold temperature of star formation	T_{th}	$1 \times 10^3\ K$	$1 \times 10^3–3 \times 10^4\ K$
Supernova explosion energy	ε_{SN}	$1 \times 10^{51}\ erg$	$(0.03 – 1) \times 10^{51}\ erg$

Notes We adopt the fiducial value of $c_\star$, n_{th}, T_{th}, ε_{SN} from Saitoh et al. [79]

where $c_\star$, $m_\star$ and m_{gas} are star formation efficiency parameter, the mass of star and gas particles, respectively (e.g., [37, 59, 93]). The ASURA sets $m_\star = m_{gas}/3$ according to Okamoto et al. [62, 63]. The mass of gas particles is reduced by the star formation. The gas particles are converted into collisionless particles if mass of them becomes less than one-third of their initial mass. The dimensionless star formation efficiency parameter of our fiducial model is set to be $c_\star = 0.033$. This value is motivated from the slow star formation models presented in Zuckerman and Evans [113], Krumholz and Tan [43]. The value of $c_\star$ does not strongly affect the properties of galaxies when we adopt $n_{th} = 100\ cm^{-3}$ ([79] and Chap. 3). Table 2.1 lists parameters adopted in this work.

The ASURA code treats star particles as simple stellar populations (SSPs). The SSP assumes that star particles are assemblies of stars which have the same metallicity and age. In SSP models, we assume the initial mass function (IMF) within a star particle. Several works derived the IMFs. In Chaps. 3 and 5, we adopt the Salpeter IMF [81]:

$$\frac{dN}{d\log_{10} m} \propto m^{-1.35}, \tag{2.10}$$

where m and N are the mass and number of stars, respectively. In Chaps. 4 and 7, we adopt Kroupa IMF [42]:

$$\frac{dN}{d\log_{10} m} \propto \begin{cases} m^{-0.3} & (0.1\ M_\odot < m < 0.5\ M_\odot), \\ m^{-1.3} & (0.5\ M_\odot < m < 120\ M_\odot). \end{cases} \tag{2.11}$$

In Chap. 6, we adopt Chabrier IMF [7].

$$\frac{dN}{d\log_{10} m} \propto \begin{cases} \exp\left[\dfrac{\{\log_{10}(\frac{m}{0.079})\}^2}{2(0.69)^2} \right] & (0.1\ M_\odot < m < 1\ M_\odot), \\ \\ m^{-1.3} & (1\ M_\odot < m < 100\ M_\odot). \end{cases} \tag{2.12}$$

All models in our simulations adopt the mass range of 0.1–100 $M_\odot$. Several studies suggested that fractions of massive stars are predicted to be higher in stars formed without metals (Population III stars) than those of stars with metals (e.g., [32, 92, 95]). We adopt top-heavy IMF inferred from Susa et al. [95] for Population III stars in Chaps. 4, 6, and 7.

2.2 Feedback from Massive Stars

2.2.1 Models of Core-Collapse Supernovae

Stars which have mass of $\gtrsim 8\ M_\odot$ explode as core-collapse supernovae. Current simulations of galaxy formation cannot resolve each star. We thus need to model supernovae for simulations of galaxy formation. Here we describe models implemented in ASURA. In ASURA, the following probability (p_{CCSN}) selects the star particles, which will explode within a time of Δt,

$$
p_{CCSN} = \frac{\displaystyle\int_{m(t)}^{m(t+\Delta t)} \phi(m')m'^{-1}\mathrm{d}m'}{\displaystyle\int_{m(t)}^{8M_\odot} \phi(m')m'^{-1}\mathrm{d}m'}, \tag{2.13}
$$

where $m(t)$ is the mass of stars which have lifetime of t. The number of core-collapse supernovae in a star particle is ~ 10 if the mass of a star particle is $\sim 1000M_\odot$. Many numerical simulations of galaxy formation treat core-collapse supernovae in a similar way (e.g., [10, 34, 61, 79, 86, 102]).

We assume that each core-collapse supernova (stars with $\gtrsim 8\ M_\odot$) distributes thermal energy of 10^{51} erg to the gas particles in the nearest neighbors. We also assume that some fraction (f_{HN}) of stars with $\gtrsim 20\ M_\odot$ explode as hypernovae. They produce thermal energy of 10 times higher than that of normal core-collapse supernovae. The assumed values of f_{HN} are 0, 0.05, and 0.5. The observations of long gamma-ray bursts suggest that $f_{HN} \approx 0.05$ [26, 66]. On the other hand, chemical evolution models of Kobayashi et al. [39] predicted that $f_{HN} = 0.5$. We adopt the Chemical Evolution Library (CELIB) to compute stellar feedback and chemical evolution [73, 74].

According to Saitoh and Makino [76], when all SSP particles distribute thermal energy to surrounding gas particles, the temperature of the region heated by a supernova is $T_{SN} \sim 10^6$ K. Typical cooling time in the region is $\sim 10^3$ yr. This temperature is much lower than the temperature heated by supernova ejecta in a star-forming region in reality where cooling time is also much longer than simulated supernova heated region. To prevent overcooling in the heated region, we have implemented a stochastic supernova feedback model but can preserve the total energy. When the ith SSP particle explodes as a supernova, the increase in temperature is

$$T_i = \frac{2\mu m_p}{3k_B} \frac{\varepsilon_{SN} E_{SN} m_{\star,i}}{N_{NB} m_{gas,i}}, \tag{2.14}$$

where μ, k_B, m_p, ε_{SN}, E_{SN}, $m_{\star,i}$, N_{NB}, and $m_{gas,i}$ are the mean molecular weight, the Boltzmann constant, the proton mass, mass fraction of stars which explode as supernovae per 1 $M_\odot$, energy ejected from a supernova, stellar mass of the ith star particle, the number of nearest neighbor particle which distribute energy from supernovae, and gas mass of the ith gas particle. In this model, we set the probability, $p_{SN,i} = T_i/T_{SN,th}$. If the random number between 0 to 1 is less than p_{SN}, the thermal energy ejected from the ith star particle is $E_{SN,i} = E_{SN,th,i}$, where

$$E_{SN,i} = \frac{3k_B T_{SN,th}}{2\mu m_p} N_{NB} m_{gas,i}. \tag{2.15}$$

If the random number is larger than p_{SN}, we set $E_{SN,i} = 0$. In this study, we assume $T_{SN,th} = 5 \times 10^7$ K. This model can increase enough temperature to prevent over cooling problem. We adopt the yields of core-collapse supernovae shown in Nomoto et al. [60].

We also put the effect of heating by H_{II} region around young massive stars. By using PÉGASE [22], we evaluate the number of the Lyα photons around these stars.

2.2.2 Models of Electron-Capture Supernovae

We newly implement the models of electron-capture supernovae in this work. The nucleosynthesis yields are taken from model e8.8 in Wanajo et al. [107]. We distribute the thermal energy of 9×10^{49} erg to the neighbor gas particles. The value of thermal energy is taken from Wanajo et al. [107].

The mass ranges of progenitors of electron-capture supernovae are highly uncertain. They are affected by the parameters of stellar evolution such as treatment of mass loss, overshooting, and dredge-up. It is, therefore, necessary to examine the effects of the mass ranges.

We adopt the mass ranges of progenitors of electron-capture supernovae from different stellar evolution models (metallicity dependent mass ranges of [16, 67]), and a constant mass range from 8.5 to 9.0 $M_\odot$. Doherty et al. [16] have shown that the mass ranges of progenitors of electron-capture supernovae are from 8.2 to 8.4 $M_\odot$ and 9.8 to 9.9 $M_\odot$ at $Z = 10^{-4}$ and 0.02, respectively. At lower metallicity, the efficiency of the CNO cycle decreases, resulting in the more massive CO cores. This effect makes the progenitor mass of electron-capture supernovae less massive in lower metallicity. Poelarends [67] shows wider mass ranges than those of Doherty et al. [16]. This difference mainly comes from the adopted mass-loss rates. Table 2.2 shows the mass ranges of three models in each metallicity.

Table 2.2 The mass ranges of progenitors of electron-capture supernovae

Z	Constant mass range		Doherty et al. [16]		Poelarends [67]	
	M_l	M_u	M_l	M_u	M_l	M_u
	$M_\odot$	$M_\odot$	$M_\odot$	$M_\odot$	$M_\odot$	$M_\odot$
1×10^{-5}	8.5	9.0	8.2	8.4	6.4	8.2
1×10^{-4}	8.5	9.0	8.2	8.4	6.9	8.2
1×10^{-3}	8.5	9.0	8.3	8.4	7.6	8.4
4×10^{-3}	8.5	9.0	8.8	9.0	8.4	9.1
8×10^{-3}	8.5	9.0	9.5	9.6	8.7	9.3
2×10^{-2}	8.5	9.0	9.8	9.9	9.0	9.3

Columns from left to right represent the metallicity (Z), the lower (M_l) and upper (M_u) mass of progenitors of electron-capture supernovae

2.2.3 Models of Neutron Star Mergers

Here we describe models of neutron star mergers newly implemented in this work. Neutron star mergers are one of the most promising sites of the r-process (Sect. 1.1.3.2). In this work, we assume that the r-process elements are synthesized by neutron star mergers.

Merger timescales of neutron star mergers are one of the key parameters to clarify the enrichment of the r-process. The merger times of neutron star mergers have been estimated by the star formation histories of a host galaxy of a kilonova/macronova, binary pulsars in the Milky Way, short gamma-ray bursts, and population synthesis calculations. Blanchard et al. [4] estimated that the merger timescale of the neutron star merger, GW170817 is $11.2^{+0.7}_{-1.4}$ billion years from the star formation histories of the host galaxy, NGC 4993. The shortest merger timescale ever found in the observed binary pulsar is 46 million years [94]. Observations of short gamma-ray bursts (e.g., [23, 48, 64, 112]) and population synthesis calculations (e.g., [9, 17]) suggested that the index of the power law of the merger time distribution of neutron star mergers was $\gtrsim -1$. Following these observations, we set merger times as 10, 100, or 500 million years in Chap. 5 and the power law merger time distributions with the index of -1 in Chaps. 6 and 7. We set minimum merger timescale to be 10^7 yr in Chaps. 6 and 7.

The merger times and the rates of neutron star mergers are highly uncertain. We vary these parameters for ~ 2 dex in this work. We assume a number fraction of neutron star mergers to the stars from 8 to 20 $M_\odot$ as 0.01 in our fiducial model. The upper mass of progenitors of neutron star mergers (20 $M_\odot$) is taken from the lower mass limit of a formation of black holes [17]. The corresponding rate of neutron star merger in a galaxy with a size of the Milky Way is $\sim 5 \times 10^{-5}$ yr^{-1}. It is within the values of the Galactic disk $\sim 10^{-6}$–10^{-3} yr^{-1}, estimated from observed compact binaries [1].

The ejecta from neutron star mergers are distributed within an SPH kernel. This is the same way used for the ejecta of supernovae. Montes et al. [58] have shown that the ejecta from neutron star mergers expand similarly with supernovae. For yields of neutron star mergers (Chaps. 6 and 7), each neutron star merger is assumed to eject $1.8 \times 10^{-4} \, M_\odot$ and $2.0 \times 10^{-5} \, M_\odot$ of Ba and Eu, respectively [106] in Chaps. 6 and 7. In Chap. 5, we adopt the empirical yield related to the rate of neutron star mergers following Ishimaru et al. [35]. We set [Eu/Fe] to be 0.5 at [Fe/H] $= 0$ without the contribution of type Ia supernovae.

Binary black hole-neutron star mergers may also eject r-process elements (e.g., [14, 21, 44, 46, 47]). The rate of these events are expected to be ~ 10 times lower than that of neutron star mergers [1]. We thus do not consider the effect of black hole-neutron star mergers in this work.

2.3 Feedback Models of Intermediate and Low Mass Stars

In this work, we adopt the empirical distribution of delay times of type Ia supernovae presented in Maoz and Mannucci [54]. The number of type Ia supernovae (N_{SNIa}) per unit time can be written as

$$\frac{\mathrm{d}N_{\mathrm{SNIa}}}{\mathrm{d}t} = \varepsilon_{\mathrm{SNIa}} \left(\frac{t}{10^9 \mathrm{yr}} \right)^{\alpha_{\mathrm{SNIa}}}, \tag{2.16}$$

where $\varepsilon_{\mathrm{SNIa}} = 4 \times 10^{-13}$ and $\alpha_{\mathrm{SNIa}} = -1$ following Maoz and Mannucci [54] to be consistent with the observed rates and delay times of type Ia supernovae. The minimum delay times of type Ia supernovae are set to be 10^8 yr according to the observed estimate of Totani et al. [100]. The ASURA distributes the energy from type Ia supernovae in the same way as core-collapse supernovae. The model N100 of Seitenzahl et al. [83] is adopted for the nucleosynthesis yields of type Ia supernovae. Following the yield, each type Ia supernova is assumed to produce $0.74 \, M_\odot$ of Fe.

The effects of asymptotic giant branch stars are also implemented. We assume a mass loss of asymptotic giant branch stars occurs within a time interval of 3×10^8 yr. We adopt the yields computed by Karakas [36].

2.4 Nucleosynthesis Yields of Supernovae

In this section, we show nucleosynthesis yields adopted in this study. For electron-capture supernovae, we adopt the yield of Wanajo et al. [107] computed in the solar metallicity. This model is adopted for all ranges of metallicity. The initial compositions are insensitive to the yields of electron-capture supernovae [105, 107].

For iron core-collapse supernovae and hypernovae, we adopt the metallicity dependent yields of Nomoto et al. [60]. Section 4.5 shows the effects of yields on the enrichment of Zn. In this section, we also adopt the yields of Chieffi and Limongi

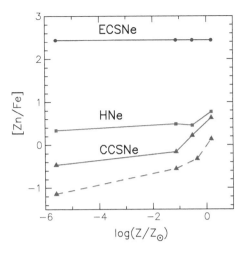

Fig. 2.1 The ratios of [Zn/Fe] in the IMF integrated yields of electron-capture supernovae (blue circles), hypernovae (green squares), and core-collapse supernovae (red triangles) as functions of metallicity (Z, Fig. 2 of Hirai et al. [31], reproduced under the terms of Creative Commons Attribution 3.0 license (https://creativecommons.org/licenses/by/3.0/)). We adopt the electron-capture supernova yields of Wanajo et al. [107]. The yields of hypernovae and core-collapse supernovae are taken from Nomoto et al. [60]. The core-collapse supernova yield of Chieffi and Limongi [8] is plotted with the red dashed lines. At $\log_{10}(Z/Z_\odot) = -5.6$, yields for zero metallicity are plotted. Metallicity of $Z_\odot = 0.0134$ is adopted for solar metallicity [2]

([8], $Z \geq 10^{-5}Z_\odot$) and Limongi and Chieffi ([49], $Z < 10^{-5}Z_\odot$). Table 2.3 lists the nucleosynthesis yields of electron-capture supernovae [107], core-collapse supernovae, and hypernovae [60].

Figure 2.1 compares yields of [Zn/Fe] in each metallicity integrated by the initial mass function. As shown in this figure, electron-capture supernovae synthesize higher ratios of [Zn/Fe] than those of iron core-collapse supernovae and hypernovae. Compared to iron core-collapse supernovae and hypernovae, electron-capture supernovae produce a much smaller amount of Fe but synthesize a large amount of Zn. The yields of electron-capture supernovae, therefore, have high [Zn/Fe] ratios. In the case of core-collapse supernovae, the assumption of the boundary between the remnant core and ejected material is critical to determine the yield of Zn. Section 4.5 shows the effects of yields of supernovae on the enrichment of Zn.

The yields of isotopes are different among electron-capture supernovae, iron core-collapse supernovae, and hypernovae. Electron-capture supernovae synthesize all stable isotopes of Zn from the neutron-rich ejecta. On the other hand, ^{64}Zn is predominantly synthesized in high entropy ejecta in hypernovae. At higher metallicity, the weak s-process synthesizes ^{66,67,68,70}Zn [39].

Figure 2.2 represents production factors, which are the mass fractions relative to the solar composition, of Zn isotopes. At zero metallicity, core-collapse supernovae and hypernovae synthesize tiny amount of ^{66,67,68,70}Zn. The effects of the weak s-process can be seen in the increase of production factors of ^{66,67,68,70}Zn at the solar metallicity.

Table 2.3 Nucleosynthesis yields of supernovae

Elements	Electron-capture supernovae	Core-collapse supernovae					Hypernovae			
	$8.8\ M_\odot$	$15\ M_\odot$	$20\ M_\odot$	$25\ M_\odot$	$30\ M_\odot$	$40\ M_\odot$	$20\ M_\odot$	$25\ M_\odot$	$30\ M_\odot$	$40\ M_\odot$
$Z = 0$										
Mg	8.1×10^{-8}	6.9×10^{-2}	1.5×10^{-1}	1.2×10^{-1}	2.3×10^{-1}	4.8×10^{-1}	1.7×10^{-1}	1.5×10^{-1}	2.2×10^{-1}	3.4×10^{-1}
Fe	3.1×10^{-3}	7.2×10^{-2}	7.2×10^{-2}	7.4×10^{-2}	7.5×10^{-2}	8.0×10^{-2}	8.5×10^{-2}	9.9×10^{-2}	1.6×10^{-1}	2.6×10^{-1}
Zn	1.1×10^{-3}	1.2×10^{-4}	8.3×10^{-5}	2.6×10^{-6}	3.1×10^{-10}	4.4×10^{-11}	3.9×10^{-4}	2.7×10^{-4}	5.9×10^{-4}	7.0×10^{-4}
$Z = 1 \times 10^{-3}$										
Mg	8.1×10^{-8}	6.5×10^{-2}	2.5×10^{-1}	1.8×10^{-1}	2.9×10^{-1}	7.1×10^{-1}	2.3×10^{-1}	2.0×10^{-1}	3.2×10^{-1}	5.2×10^{-1}
Fe	3.1×10^{-3}	7.2×10^{-2}	7.2×10^{-2}	7.2×10^{-2}	7.3×10^{-2}	7.9×10^{-2}	8.2×10^{-2}	1.5×10^{-1}	2.0×10^{-1}	2.6×10^{-1}
Zn	1.1×10^{-3}	6.8×10^{-5}	3.3×10^{-5}	4.9×10^{-5}	8.2×10^{-5}	1.1×10^{-4}	3.6×10^{-4}	5.7×10^{-4}	8.5×10^{-4}	7.3×10^{-4}
$Z = 4 \times 10^{-3}$										
Mg	8.1×10^{-8}	7.7×10^{-2}	9.8×10^{-2}	2.4×10^{-1}	2.3×10^{-1}	4.0×10^{-1}	8.2×10^{-2}	2.3×10^{-1}	2.0×10^{-1}	4.2×10^{-1}
Fe	3.1×10^{-3}	7.0×10^{-2}	7.1×10^{-2}	6.9×10^{-2}	7.3×10^{-2}	7.4×10^{-2}	2.6×10^{-2}	7.8×10^{-2}	1.5×10^{-1}	2.7×10^{-1}
Zn	1.1×10^{-3}	1.2×10^{-4}	8.5×10^{-5}	2.2×10^{-4}	2.6×10^{-4}	5.7×10^{-4}	6.0×10^{-5}	4.0×10^{-4}	5.4×10^{-4}	1.0×10^{-3}
$Z = 2 \times 10^{-2}$										
Mg	8.1×10^{-8}	3.3×10^{-2}	8.4×10^{-2}	2.5×10^{-1}	2.3×10^{-1}	4.4×10^{-1}	7.6×10^{-2}	2.5×10^{-1}	2.0×10^{-1}	4.3×10^{-1}
Fe	3.1×10^{-3}	6.6×10^{-2}	6.1×10^{-2}	5.9×10^{-2}	6.1×10^{-2}	5.2×10^{-2}	9.3×10^{-3}	9.4×10^{-2}	7.3×10^{-2}	2.6×10^{-1}
Zn	1.1×10^{-3}	7.1×10^{-5}	2.3×10^{-4}	9.2×10^{-4}	7.1×10^{-6}	3.0×10^{-3}	1.6×10^{-4}	1.1×10^{-3}	1.3×10^{-4}	3.3×10^{-3}

Columns show names of elements and yields of Mg, Fe, and Zn. Note that the yield of electron-capture supernovae is computed only around the core

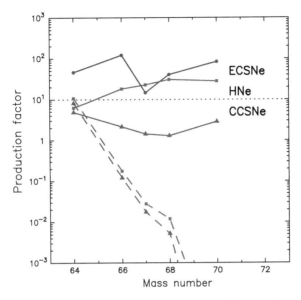

Fig. 2.2 Production factors of Zn isotopes (Fig. 3 of Hirai et al. [31], reproduced under the terms of Creative Commons Attribution 3.0 license (https://creativecommons.org/licenses/by/3.0/)). The production factors predicted by models of core-collapse supernovae (15 $M_\odot$), hypernovae (25 $M_\odot$), and electron-capture supernovae (8.8 $M_\odot$) are shown with red triangles, green squares, and blue circles, respectively. Models with solar and zero metallicity are respectively shown with solid and dashed lines. The production factor of 10 is shown with the black dotted line. Astrophysical sites above this value can be the main site of a given isotope (e.g., [110])

2.5 Ejection of Metals

We distribute metals from star particles to surrounding gas particles within SPH kernel. The initial gas metallicity is set to be zero. When a star particle ejects metals, neighbor gas particles gain these metals. The amount of metals inherited to gas particles is weighted by the distance from the star particle. We set the number of nearest neighbor gas particles to be 32 ± 2 (Chaps. 3 and 5) or 128 ± 8 (Chaps. 4, 6, and 7) as a fiducial value. For simplicity, these values are set to be the same as the number of nearest neighbor particles for SPH kernel. The increase of mass of element X ($\Delta M_{X,j}$) in the jth gas particle due to the ejection of metals from the ith star particle is defined as follows:

$$\Delta M_{X,j} = \frac{m_j}{\rho_i} M_{X,i} W(r_{ij}, h_{ij}), \tag{2.17}$$

where h_{ij}, r_{ij}, and $W(r_{ij}, h_{ij})$ are the smoothing length, the distance between particle i and j, the SPH kernel of a cubic spline function (e.g., [38], Chaps. 3 and 5) or Wendland kernel C4 ([15, 108] Chaps. 4, 6, and 7), respectively.

2.6   Mixing of Metals

2.6.1   Definition of Metallicity in Stars

Abundances of elements in star-forming regions are directly reflected the elemental abundances of newly formed stars. In this study, we define the metallicity of newly formed stars in two ways. One is to take the abundances of a gas particle which formed the star particle (e.g., [69, 85, 101]). The other way is to assume the average value of abundances in surrounding gas particles (e.g., [30, 85]). The former is widely used in simulations of galaxies. However, this method causes a serious problem when we discuss galactic chemical evolution. Wiersma et al. [109] pointed out that metal density significantly fluctuates even in the swept-up shell of supernova remnants in SPH simulations without metal mixing. We discuss this problem in Chap. 6.

When we take the average value of abundances of elements in the star-forming region, we can avoid reflecting the fluctuation of the elemental abundances of gas particles. This model is based on the observations of the homogeneous metallicity distribution within open clusters (e.g., [6, 11–13, 65, 70, 71, 99]). This homogeneity is caused by the turbulent mixing in star-forming regions [18]. The timescale of the metal mixing is less than 1 million years. On the other hand, in a slow star formation model, the typical timescale of star formation is $\gtrsim 10$ million years [43, 113]. In this study, we take an average value of metallicity in 32 particles within an SPH kernel. In our fiducial model, the mass of the metal mixing region is $\sim 10^4 M_\odot$, which is consistent with the mass of giant molecular clouds (e.g., [28, 45, 50, 82, 88]). In Sect. 6.3, we show that metal mixing erases the abundances significantly deviate from the average value.

2.6.2   Metal Mixing Models for Gas Particles

In SPH simulations, it is necessary to put the effects of metal mixing among gas particles. Wiersma et al. [109] proposed the model which evaluates the density of metals (ρ_i^X) from the neighboring particles within an SPH kernel following Eq. (2.3),

$$\rho_i^X = \sum_j m_j \frac{\rho_j^X}{\rho_j} W_{ij}(h). \qquad (2.18)$$

This method (called *smooth metallicity*) has an advantage that it does not require additional parameters for metal mixing.

The other way to implement the metal mixing is to compute metal diffusion among gas particles. The diffusion equation for metal mixing is as follows,

$$\frac{\mathrm{d}Z_i}{\mathrm{d}t} = \nabla(D\nabla Z_i), \qquad (2.19)$$

where Z_i is the amount of elements and D is the diffusion coefficient. Greif et al. [25] adopt velocity dispersion to evaluate the diffusion coefficient. The diffusion coefficient in their model (D_i) is described as follows,

$$D_i = 2\rho_i \tilde{v} h_i, \tag{2.20}$$

where ρ_i, h_i, and $\tilde{v}$ are the density and SPH smoothing length, and

$$\tilde{v}^2 = \frac{1}{N_{\text{ngb}}} \sum_j |v_i - v_j|^2, \tag{2.21}$$

respectively. In Eq. (2.21), N_{ngb} is the number of particles in the nearest neighbor.

On the other hand, we adopt a shear based mixing model in DISPH [29, 74]. This model is based on Smagorinsky [87] and Shen et al. [84]. The kth element in the ith gas particle ($Z_{k,i}$) diffuses to neighbor gas particles following

$$\frac{dZ_{k,i}}{dt} = -\sum_j \frac{m_i}{(\rho_i + \rho_j)/2} \frac{4D_i D_j}{(D_i + D_j)} \frac{(Z_{k,i} - Z_{k,j})}{|r_{ij}|^2} r_{ij} \cdot \nabla W_{ij}, \tag{2.22}$$

$$\hat{S}_{ab,i} = \frac{1}{q_i} \sum_j U_j (v_{b,i} - v_{a,j}) \nabla_a W_{ij}, \tag{2.23}$$

$$S_{ab,i} = \frac{1}{2}(\hat{S}_{ab,i} + \hat{S}_{ba,i}) - \delta_{ab} \frac{1}{3} \text{Trace } \hat{S}_{ab,i}, \tag{2.24}$$

$$D_i = C_d |S_{ab,i}| h_i^2, \tag{2.25}$$

where a and b are the direction of x, y, and z axises, δ_{ab} is the Kronecker's delta, U_i is the internal energy, and q_i is the energy density defined by

$$q_i = \sum_j U_j W_{ij}. \tag{2.26}$$

Here we changed the scaling factor for metal diffusion (C_d) to compute models with different efficiency of metal mixing. In Chap. 6, we constrain the diffusion coefficient from the abundances of heavy elements in metal-poor stars.

2.7 Summary

In this chapter, we describe the method for chemodynamical simulations of galaxies. We adopt the N-body/SPH code, ASURA. In this code, metallicity dependent cooling/heating function, star formation, and feedback from stars are adopted. We show the method to implement the feedback from iron core-collapse supernovae, electron-capture supernovae, neutron star mergers, type Ia supernovae, and asymptotic giant branch stars. We also show that nucleosynthesis yields adopted in this study. In the last section of this chapter, we describe metal mixing models for SPH simulations.

References

1. Abadie J et al (2010) TOPICAL REVIEW: predictions for the rates of compact binary coalescences observable by ground-based gravitational-wave detectors. Class Quantum Gravity 27(17):173001
2. Asplund M, Grevesse N, Sauval AJ, Scott P (2009) The chemical composition of the sun. Annu Rev Astron Astrophys 47:481–522
3. Barnes J, Hut P (1986) A hierarchical O(N log N) force-calculation algorithm. Nature 324:446–449
4. Blanchard PK, Berger E, Fong W, Nicholl M, Leja J, Conroy C, Alexander KD, Margutti R, Williams PKG, Doctor Z, Chornock R, Villar VA, Cowperthwaite PS, Annis J, Brout D, Brown DA, Chen HY, Eftekhari T, Frieman JA, Holz DE, Metzger BD, Rest A, Sako M, Soares-Santos M (2017) The electromagnetic counterpart of the binary neutron star merger LIGO/Virgo GW170817. VII. Properties of the host galaxy and constraints on the merger timescale. Astrophys J Lett 848:L22
5. Bryan GL, Norman ML, O'Shea BW, Abel T, Wise JH, Turk MJ, Reynolds DR, Collins DC, Wang P, Skillman SW, Smith B, Harkness RP, Bordner J, Kim J-H, Kuhlen M, Xu H, Goldbaum N, Hummels C, Kritsuk AG, Tasker E, Skory S, Simpson CM, Hahn O, Oishi JS, So GC, Zhao F, Cen R, Li Y, Collaboration E (2014) ENZO: an adaptive mesh refinement code for astrophysics. Astrophys J Suppl 211(19):19
6. Bubar EJ, King JR (2010) Spectroscopic abundances and membership in the wolf 630 moving group. Astron J 140:293–318
7. Chabrier G (2003) Galactic stellar and substellar initial mass function. Publ Astron Soc Pac 115:763–795
8. Chieffi A, Limongi M (2004) Explosive yields of massive stars from $Z = 0$ to $Z = Z_{solar}$. Astrophys J 608:405–410
9. Chruslinska M, Belczynski K, Klencki J, Benacquista M (2018) Double neutron stars: merger rates revisited. Mon Not R Astron Soc 474:2937–2958
10. Dalla Vecchia C, Schaye J (2012) Simulating galactic outflows with thermal supernova feedback. Mon Not R Astron Soc 426:140–158
11. De Silva GM, Freeman KC, Asplund M, Bland-Hawthorn J, Bessell MS, Collet R (2007a) Chemical homogeneity in collinder 261 and implications for chemical tagging. Astron J 133:1161–1175
12. De Silva GM, Freeman KC, Bland-Hawthorn J, Asplund M, Bessell MS (2007b) Chemically tagging the HR 1614 moving group. Astron J 133:694–704
13. De Silva GM, D'Orazi V, Melo C, Torres CAO, Gieles M, Quast GR, Sterzik M (2013) Search for associations containing young stars: chemical tagging IC 2391 and the Argus association. Mon Not R Astron Soc 431:1005–1018
14. Deaton MB, Duez MD, Foucart F, O'Connor E, Ott CD, Kidder LE, Muhlberger CD, Scheel MA, Szilagyi B (2013) Black hole-neutron star mergers with a hot nuclear equation of state: outflow and neutrino-cooled disk for a low-mass, high-spin case. Astrophys J 776(47):47
15. Dehnen W, Aly H (2012) Improving convergence in smoothed particle hydrodynamics simulations without pairing instability. Mon Not R Astron Soc 425:1068–1082
16. Doherty CL, Gil-Pons P, Siess L, Lattanzio JC, Lau HHB (2015) Super- and massive AGB stars-IV. Final fates-initial-to-final mass relation. Mon Not R Astron Soc 446:2599–2612

17. Dominik M, Belczynski K, Fryer C, Holz DE, Berti E, Bulik T, Mandel I, O'Shaughnessy R (2012) Double compact objects. I. The significance of the common envelope on merger rates. Astrophys J 759:52
18. Feng Y, Krumholz MR (2014) Early turbulent mixing as the origin of chemical homogeneity in open star clusters. Nature 513:523–525
19. Ferland GJ, Korista KT, Verner DA, Ferguson JW, Kingdon JB, Verner EM (1998) CLOUDY 90: numerical simulation of plasmas and their spectra. Publ Astron Soc Pac 110:761–778
20. Ferland GJ, Porter RL, van Hoof PAM, Williams RJR, Abel NP, Lykins ML, Shaw G, Henney WJ, Stancil PC (2013) The 2013 release of cloudy. Rev Mex de Astron y Astrofis 49:137–163
21. Fernandez R, Foucart F, Kasen D, Lippuner J, Desai D, Roberts LF (2017) Dynamics, nucleosynthesis, and kilonova signature of black hole-neutron star merger ejecta. Class Quantum Gravity 34(15):154001
22. Fioc M, Rocca-Volmerange B (1997) PEGASE: a UV to NIR spectral evolution model of galaxies. Application to the calibration of bright galaxy counts. Astron Astrophys 326:950–962
23. Fong W, Berger E, Chornock R, Margutti R, Levan AJ, Tanvir NR, Tunnicliffe RL, Czekala I, Fox DB, Perley DA, Cenko SB, Zauderer BA, Laskar T, Persson SE, Monson AJ, Kelson DD, Birk C, Murphy D, Servillat M, Anglada G (2013) Demographics of the galaxies hosting shortduration gamma-ray bursts. Astrophys J 769(56):56
24. Gingold RA, Monaghan JJ (1977) Smoothed particle hydrodynamics-Theory and application to non-spherical stars. Mon Not R Astron Soc 181:375–389
25. Greif TH, Glover SCO, Bromm V, Klessen RS (2009) Chemical mixing in smoothed particle hydrodynamics simulations. Mon Not R Astron Soc 392:1381–1387
26. Guetta D, Della Valle M (2007) On the rates of gamma-ray bursts and type Ib/c supernovae. Astrophys J Lett 657:L73–L76
27. Haardt F, Madau P (2012) Radiative transfer in a clumpy universe. IV. New synthesis models of the cosmic UV/X-Ray background. Astrophys J 746:125
28. Heyer M, Krawczyk C, Duval J, Jackson JM (2009) Re-examining Larson's scaling relationships in galactic molecular clouds. Astrophys J 699:1092–1103
29. Hirai Y, Saitoh TR (2017) Efficiency of metal mixing in dwarf galaxies. Astrophys J Lett 838:L23
30. Hirai Y, Ishimaru Y, Saitoh TR, Fujii MS, Hidaka J, Kajino T (2015) Enrichment of r-process elements in dwarf spheroidal galaxies in chemodynamical evolution model. Astrophys J 814:41
31. Hirai Y, Saitoh TR, Ishimaru Y, Wanajo S (2018) Enrichment of zinc in galactic chemodynamical evolution models. Astrophys J 855(63):63
32. Hirano S, Hosokawa T, Yoshida N, Omukai K, Yorke HW (2015) Primordial star formation under the influence of far ultraviolet radiation: 1540 cosmological haloes and the stellar mass distribution. Mon Not R Astron Soc 448:568–587
33. Hopkins PF (2013) A general class of Lagrangian smoothed particle hydrodynamics methods and implications for fluid mixing problems. Mon Not R Astron Soc 428:2840–2856
34. Hopkins PF, Kereš D, Oñorbe J, Faucher-Giguère C-A, Quataert E, Murray N, Bullock JS, (2014) Galaxies on FIRE (Feedback In Realistic Environments): stellar feedback explains cosmologically inefficient star formation. Mon Not R Astron Soc 445:581–603
35. Ishimaru Y, Wanajo S, Prantzos N (2015) Neutron star mergers as the origin of r-process elements in the galactic halo based on the sub-halo clustering scenario. Astrophys J Lett 804:L35
36. Karakas AI (2010) Updated stellar yields from asymptotic giant branch models. Mon Not R Astron Soc 403:1413–1425
37. Katz N, Weinberg DH, Hernquist L (1996) Cosmological simulations with TreeSPH. Astrophys J 105:19
38. Kawata D (2001) Effects of Type II and Type Ia supernovae feedback on the chemodynamical evolution of elliptical galaxies. Astrophys J 558:598–614

39. Kobayashi C, Umeda H, Nomoto K, Tominaga N, Ohkubo T (2006) Galactic chemical evolution: carbon through zinc. Astrophys J 653:1145–1171
40. Kravtsov AV (1999) High-resolution simulations of structure formation in the universe. PhD thesis. New Mexico State University
41. Kravtsov AV, Klypin AA, Khokhlov AM (1997) Adaptive refinement tree: a new high-resolution N-body code for cosmological simulations. Astrophys J Suppl Ser 111:73–94
42. Kroupa P (2001) On the variation of the initial mass function. Mon Not R Astron Soc 322:231–246
43. Krumholz MR, Tan JC (2007) Slow star formation in dense gas: evidence and implications. Astrophys J 654:304–315
44. Kyutoku K, Kiuchi K, Sekiguchi Y, Shibata M, Taniguchi K (2018) Neutrino transport in black hole-neutron star binaries: neutrino emission and dynamical mass ejection. Phys Rev D 97(2):023009
45. Larson RB (1981) Turbulence and star formation in molecular clouds. Mon Not R Astron Soc 194:809–826
46. Lattimer JM, Schramm DN (1974) Black-hole-neutron-star collisions. Astrophys J Lett 192:L145–L147
47. Lattimer JM, Schramm DN (1976) The tidal disruption of neutron stars by black holes in close binaries. Astrophys J 210:549–567
48. Leibler CN, Berger E (2010) The stellar ages and masses of short gammaray burst host galaxies: investigating the progenitor delay time distribution and the role of mass and star formation in the short gamma-ray burst rate. Astrophys J 725:1202–1214
49. Limongi M, Chieffi A (2012) Presupernova evolution and explosive nucleosynthesis of zero metal massive stars. Astrophys J Suppl 199:38
50. Liszt HS, Delin X, Burton WB (1981) Properties of the galactic molecular cloud ensemble from observations of /C-13/O. Astrophys J 249:532–549
51. Lucy LB (1977) A numerical approach to the testing of the fission hypothesis. Astrophys J 82:1013–1024
52. Makino J (1991) A modified aarseth code for GRAPE and vector processors. Publ Astron Soc Jpn 43:859–876
53. Makino J (2004) A fast parallel treecode with GRAPE. Publ Astron Soc Jpn 56:521–531
54. Maoz D, Mannucci F (2012) Type-Ia supernova rates and the progenitor problem: a review. Publ Astron Soc Aust 29:447–465
55. McMillan SLW (1986) The vectorization of small-N integrators. In: Hut P, McMillan SLW (eds) The use of supercomputers in stellar dynamics, vol 267. Lecture notes in physics, Springer, Berlin, p 156
56. Monaghan JJ (1992) Smoothed particle hydrodynamics. Annu Rev Astron Astrophys 30:543–574
57. Monaghan JJ, Lattanzio JC (1985) A refined particle method for astrophysical problems. Astron Astrophys 149:135–143
58. Montes G, Ramirez-Ruiz E, Naiman J, Shen S, Lee WH (2016) Transport and mixing of r-process elements in neutron star binary merger blast waves. Astrophys J 830:12
59. Navarro JF, White SDM (1993) Simulations of dissipative galaxy formation in hierarchically clustering universes-Part one-Tests of the code. Mon Not R Astron Soc 265:271
60. Nomoto K, Kobayashi C, Tominaga N (2013) Nucleosynthesis in stars and the chemical enrichment of galaxies. Annu Rev Astron Astrophys 51:457–509
61. Okamoto T, Nemmen RS, Bower RG (2008) The impact of radio feedback from active galactic nuclei in cosmological simulations: formation of disc galaxies. Mon Not R Astron Soc 385:161–180
62. Okamoto T, Jenkins A, Eke VR, Quilis V, Frenk CS (2003) Momentum transfer across shear flows in smoothed particle hydrodynamic simulations of galaxy formation. Mon Not R Astron Soc 345:429–446
63. Okamoto T, Eke VR, Frenk CS, Jenkins A (2005) Effects of feedback on the morphology of galaxy discs. Mon Not R Astron Soc 363:1299–1314

64. O'Shaughnessy R, Belczynski K, Kalogera V (2008) Short gamma-ray bursts and binary mergers in spiral and elliptical galaxies: redshift distribution and hosts. Astrophys J 675:566–585

65. Pancino E, Carrera R, Rossetti E, Gallart C (2010) Chemical abundance analysis of the open clusters Cr 110, NGC 2099 (M 37), NGC 2420, NGC 7789, and M 67 (NGC 2682). Astron Astrophys 511:A56

66. Podsiadlowski P, Mazzali PA, Nomoto K, Lazzati D, Cappellaro E (2004) The rates of hypernovae and gamma-ray bursts: implications for their progenitors. Astrophys J Lett 607:L17–L20

67. Poelarends AJT (2007) Stellar evolution on the borderline of white dwarf and neutron star formation. PhD thesis. Utrecht University

68. Rahmati A, Pawlik AH, M. Raičeviè, Schaye J, (2013) On the evolution of the H I column density distribution in cosmological simulations. Mon Not R Astron Soc 430:2427–2445

69. Raiteri CM, Villata M, Gallino R, Busso M, Cravanzola A (1999) Simulations of galactic chemical evolution: BA enrichment. Astrophys J Lett 518:L91–L94

70. Reddy ABS, Giridhar S, Lambert DL (2012) Comprehensive abundance analysis of red giants in the open clusters NGC 752, 1817, 2360 and 2506. Mon Not R Astron Soc 419:1350–1361

71. Reddy ABS, Giridhar S, Lambert DL (2013) Comprehensive abundance analysis of red giants in the open clusters NGC 2527, 2682, 2482, 2539, 2335, 2251 and 2266. Mon Not R Astron Soc 431:3338–3348

72. Rudd DH, Zentner AR, Kravtsov AV (2008) Effects of baryons and dissipation on the matter power spectrum. Astrophys J 672(19–32):19–32

73. Saitoh TR (2016) CELib: software library for simulations of chemical evolution. Astrophysics Source Code Library

74. Saitoh TR (2017) Chemical evolution library for galaxy formation simulation. Astrophys J 153:85

75. Saitoh TR, Makino J (2009) A necessary condition for individual time steps in SPH simulations. Astrophys J Lett 697:L99–L102

76. Saitoh TR, Makino J (2010) FAST: a fully asynchronous split time-integrator for a self-gravitating fluid. Publ Astron Soc Jpn 62:301–314

77. Saitoh TR, Makino J (2013) A density-independent formulation of smoothed particle hydrodynamics. Astrophys J 768:44

78. Saitoh TR, Makino J (2016) Santa barbara cluster comparison test with DISPH. Astrophys J 823:144

79. Saitoh TR, Daisaka H, Kokubo E, Makino J, Okamoto T, Tomisaka K, Wada K, Yoshida N (2008) Toward first-principle simulations of galaxy formation: I. How should we choose star-formation criteria in high-resolution simulations of disk galaxies? Publ Astron Soc Jpn 60:667–681

80. Saitoh TR, Daisaka H, Kokubo E, Makino J, Okamoto T, Tomisaka K, Wada K, Yoshida N (2009) Toward first-principle simulations of galaxy formation: II. Shock-induced starburst at a collision interface during the first encounter of interacting galaxies. Publ Astron Soc Jpn 61:481–486

81. Salpeter EE (1955) The luminosity function and stellar evolution. Astrophys J 121:161

82. Sanders DB, Scoville NZ, Solomon PM (1985) Giant molecular clouds in the Galaxy. II-Characteristics of discrete features. Astrophys J 289:373–387

83. Seitenzahl IR, Ciaraldi-Schoolmann F, Ropke FK, Fink M, Hillebrandt W, Kromer M, Pakmor R, Ruiter AJ, Sim SA, Taubenberger S (2013) Three-dimensional delayed-detonation models with nucleosynthesis for Type Ia supernovae. Mon Not R Astron Soc 429:1156–1172

84. Shen S, Wadsley J, Stinson G (2010) The enrichment of the intergalactic medium with adiabatic feedback-I. Metal cooling and metal diffusion. Mon Not R Astron Soc 407:1581–1596

85. Shen S, Cooke RJ, Ramirez-Ruiz E, Madau P, Mayer L, Guedes J (2015) The history of R-process enrichment in the Milky Way. Astrophys J 807:115

86. Simpson CM, Bryan GL, Johnston KV, Smith BD, Mac Low M-M, Sharma S, Tumlinson J (2013) The effect of feedback and reionization on star formation in low-mass dwarf galaxy haloes. Mon Not R Astron Soc 432:1989–2011

87. Smagorinsky J (1963) General circulation experiments with the primitive equations. Mon Weather Rev 91:99
88. Solomon PM, Rivolo AR, Barrett J, Yahil A (1987) Mass, luminosity, and line width relations of Galactic molecular clouds. Astrophys J 319:730–741
89. Springel V (2005) The cosmological simulation code GADGET-2. Mon Not R Astron Soc 364:1105–1134
90. Springel V (2010) E pur si muove: Galilean-invariant cosmological hydrodynamical simulations on a moving mesh. Mon Not R Astron Soc 401:791–851
91. Springel V, Yoshida N, White SDM (2001) GADGET: a code for collisionless and gasdynamical cosmological simulations. New Astron 6:79–117
92. Stacy A, Bromm V, Lee AT (2016) Building up the Population III initial mass function from cosmological initial conditions. Mon Not R Astron Soc 462:1307–1328
93. Stinson G, Seth A, Katz N, Wadsley J, Governato F, Quinn T (2006) Star formation and feedback in smoothed particle hydrodynamic simulations-I. Isolated galaxies. Mon Not R Astron Soc 373:1074–1090
94. Stovall K et al (2018) PALFA discovery of a highly relativistic double neutron star binary. Astrophys J 854(L22):L22
95. Susa H, Hasegawa K, Tominaga N (2014) The mass spectrum of the first stars. Astrophys J 792(32):32
96. Tanikawa A, Yoshikawa K, Okamoto T, Nitadori K (2012) N-body simulation for self-gravitating collisional systems with a new SIMD instruction set extension to the x86 architecture, Advanced Vector eXtensions. New Astron 17:82–92
97. Tanikawa A, Yoshikawa K, Nitadori K, Okamoto T (2013) Phantom-GRAPE: numerical software library to accelerate collisionless N-body simulation with SIMD instruction set on x86 architecture. New Astron 19:74–88
98. Teyssier R (2002) Cosmological hydrodynamics with adaptive mesh refinement. A new high resolution code called RAMSES. Astron Astrophys 385:337–364
99. Ting Y-S, De Silva GM, Freeman KC, Parker SJ (2012) Highresolution elemental abundance analysis of the open cluster IC 4756. Mon Not R Astron Soc 427:882–892
100. Totani T, Morokuma T, Oda T, Doi M, Yasuda N (2008) Delay time distribution measurement of Type Ia supernovae by the Subaru/XMM-newton deep survey and implications for the progenitor. Publ Astron Soc Jpn 60:1327–1346
101. van de Voort F, Quataert E, Hopkins PF, D. Kereš, Faucher-Giguère C-A, (2015) Galactic r-process enrichment by neutron star mergers in cosmological simulations of a Milky Way-mass galaxy. Mon Not R Astron Soc 447:140–148
102. Vogelsberger M, Genel S, Sijacki D, Torrey P, Springel V, Hernquist L (2013) A model for cosmological simulations of galaxy formation physics. Mon Not R Astron Soc 436:3031–3067
103. Wadsley JW, Keller BW, Quinn TR (2017) Gasoline2: a modern smoothed particle hydrodynamics code. Mon Not R Astron Soc 471:2357–2369
104. Wadsley JW, Stadel J, Quinn T (2004) Gasoline: a flexible, parallel implementation of TreeSPH. New Astron 9:137–158
105. Wanajo S, Janka H-T, Muller B (2011) Electron-capture supernovae as the origin of elements beyond iron. Astrophys J Lett 726:L15
106. Wanajo S, Sekiguchi Y, Nishimura N, Kiuchi K, Kyutoku K, Shibata M (2014) Production of all the r-process nuclides in the dynamical ejecta of neutron star mergers. Astrophys J Lett 789:L39
107. Wanajo S, Muller B, Janka H-T, Heger A (2018) Nucleosynthesis in the innermost ejecta of neutrino-driven supernova explosions in two dimensions. Astrophys J 852:40
108. Wendland H (1995) Piecewise polynomial, positive definite and compactly supported radial functions of minimal degree. Adv Comput Math 4(1):389–396. ISSN: 1572-9044
109. Wiersma RPC, Schaye J, Theuns T, Dalla Vecchia C, Tornatore L (2009) Chemical enrichment in cosmological, smoothed particle hydrodynamics simulations. Mon Not R Astron Soc 399:574–600

110. Woosley SE, Heger A (2007) Nucleosynthesis and remnants in massive stars of solar metallicity. Phys Rep 442:269–283

111. Yamamoto S, Saitoh TR, Makino J (2015) Smoothed particle hydrodynamics with smoothed pseudo-density. Publ Astron Soc Jpn 67:37

112. Zheng Z, Ramirez-Ruiz E (2007) Deducing the lifetime of short gamma-ray burst progenitors from host galaxy demography. Astrophys J 665:1220–1226

113. Zuckerman B, Evans NJ II (1974) Models of massive molecular clouds. Astrophys J Lett 192:L149–L152

Chapter 3
Chemodynamical Evolution of Dwarf Galaxies

Abstract Abundances of elements in metal-poor stars reflect the evolutionary histories of galaxies. Chemical evolution of galaxies is highly related to dynamical evolution with them. It is necessary to conduct chemodynamical studies to fully understand the evolutionary histories of galaxies. This chapter aims to show that the isolated dwarf galaxy models used in this work can reproduce basic observed properties of dwarf galaxies in the Local Group. In this chapter, we compare computed radial profiles, star formation histories, metallicity distribution functions, mass-metallicity relations, and α-element abundances with observed ones. We also discuss the parameter dependence on these quantities (Contents in this chapter have been in part published in [15–17] reproduced by permission of the AAS and Oxford University Press).

3.1 Models of Dwarf Spheroidal Galaxies

There are two ways to put initial conditions for simulations of galaxies. One way is to adopt models of galaxies (e.g., [1, 12, 19, 21, 29, 37]). We can reduce computation cost by adopting isolated galaxy models if we focus on the studies of specific region or objects.

On the other hand, it cannot follow the evolutionary histories of galaxies fully. The other way is to simulate galaxies with cosmological initial conditions (e.g., [20, 34, 40]). It can follow the accretion and merging of halos. However, the computational cost is usually expensive to conduct such simulations. In this book, we simulate galaxies with isolated initial conditions from this chapter to Chap. 6. In Chap. 7, we discuss the enrichment of heavy elements in cosmological initial conditions.

Here we describe the models of isolated dwarf galaxies adopted in this study. Observations of the density profiles of dwarf galaxies suggest that the shape of the density profile is not cusped but a cored at the center of halos (e.g., [32, 33]). Based on models of Revaz et al. [38] and Revaz and Jablonka [37], we assume a pseudo isothermal profile [2] for both gas and dark matter as an initial density profile:

$$\rho = \frac{\rho_c}{1 + (r/r_c)^2},$$

(3.1)

© Springer Nature Singapore Pte Ltd. 2019
Y. Hirai, *Understanding the Enrichment of Heavy Elements
by the Chemodynamical Evolution Models of Dwarf Galaxies*,
Springer Theses, https://doi.org/10.1007/978-981-13-7884-3_3

where r_c and ρ_c represent the core radius and the central density. Total mass in this system (M_{tot}) satisfies the following equation,

$$M_{tot} = 4\pi\rho_c r_c^3 \left[\frac{r_{max}}{r_c} - \arctan\left(\frac{r_{max}}{r_c}\right) \right], \tag{3.2}$$

where r_{max} is the maximum outer radius. We also assume initial velocity dispersions. For dark matter particles, we adopt isotopic velocity dispersions [3],

$$\sigma^2(r) = \frac{1}{\rho(r)} \int_r^\infty dr' \rho(r') \frac{\partial \Phi(r')}{\partial r'}, \tag{3.3}$$

where $\Phi(r)$ denotes the gravitational potential. We set initial velocity dispersion to be 0 and initial temperature for 10^4 K for gas particles. Table 3.1 lists models adopted in this chapter. We fix $r_{max} = 7.1\, r_c$ in the fiducial model. In model D, we set $r_{max} = r_c = 1$ kpc to make a model with very high density.

We varied the total mass and central density in our models. Several studies have shown that the initial density and the total mass of halos significantly affect the final properties of dwarf galaxies (e.g., [4, 5, 37, 38, 47]). Carraro et al. [4] showed that the star formation histories are affected by the initial density. Valcke et al. [47] reproduced various observational properties such as surface brightness profiles, central velocity dispersion, and luminosity-metallicity relations in dwarf galaxies with different total

Table 3.1 List of our models discussed in this chapter

Model	M_{tot} $10^8 M_\odot$	ρ_c $10^7 M_\odot$ kpc^{-3}	m_{DM} $10^3 M_\odot$	m_{gas} $10^3 M_\odot$	r_{max} kpc	r_c kpc	n_{th} cm^{-3}	$c_\star$	ε_{SN} erg
A	3.5	0.5	1.1	0.2	7.1	1.0	100	0.033	1.0×10^{51}
B	3.5	1.5	1.1	0.2	4.9	0.7	100	0.033	1.0×10^{51}
C	3.5	10.0	1.1	0.2	2.6	0.4	100	0.033	1.0×10^{51}
D	3.5	10.0	1.1	0.2	1.0	1.0	100	0.033	1.0×10^{51}
E	7.0	0.5	2.3	0.4	8.9	1.3	100	0.033	1.0×10^{51}
F	35.0	0.5	11.3	2.0	15.3	2.2	100	0.033	1.0×10^{51}
G	7.0	0.5	4.5	0.8	8.9	1.3	100	0.033	1.0×10^{51}
Q	7.0	0.5	0.8	0.8	8.9	1.3	100	0.033	1.0×10^{51}
s000	7.0	0.5	2.3	0.4	7.1	1.0	100	0.033	1.0×10^{51}
sn10	7.0	0.5	2.3	0.4	7.1	1.0	10	0.033	1.0×10^{51}
sc50	7.0	0.5	2.3	0.4	7.1	1.0	100	0.5	1.0×10^{51}
se01	7.0	0.5	2.3	0.4	7.1	1.0	100	0.033	1.0×10^{50}
mExt	7.0	0.5	2.3	0.4	7.1	1.0	0.1	0.033	3.0×10^{49}

Columns from left to right represent names of models, the total mass of halos (M_{tot}), the initial central density of halos (ρ_c), the mass of a dark matter particle (m_{DM}), the mass of a gas particle (m_{gas}), the maximum radius of halos (r_{max}), the core radius of halos (r_c), the threshold density for star formation (n_{th}), the dimensionless star formation efficiency ($c_\star$), and the supernova feedback energy (ε_{SN})

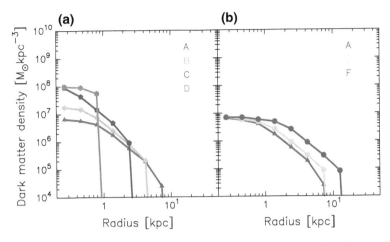

Fig. 3.1 The Initial dark matter density profiles of halos (Fig. 1 of [16] reproduced by permission of the Oxford University Press). Panel **a** shows the density profiles of models A (red), B (green), C (blue), and D (orange). Panel **b** represents models A (red), E (cyan), and F (magenta)

mass. Revaz and Jablonka [37] pointed out that the central density and total mass of halos significantly affect the stellar mass and metallicity.

Figure 3.1 denotes the initial radial dark matter density profiles of our models. Figure 3.1a represents models with different central density. The maximum radius is significantly varied to fix the initial total mass to be $3.5 \times 10^8 \ M_\odot$. Figure 3.1b shows models with different total mass. As shown in this figure, the central density is constant in these models.

3.2 Dynamical Evolution of Dwarf Galaxies

This section shows the dynamical evolution of dwarf galaxy models. Figure 3.2 shows snapshots of gas and stellar surface densities. Upper panel denotes gas surface densities. In this model, gases fall to the center of the halo following the collapse of dark matter halo initially assumed not in dynamical equilibrium and radiative cooling. Red region in this figure denotes the star-forming region. As time passes, the amount of gas decreases because star formation consumes gas. Heating of gas by supernova feedback also contributes to reducing the amount of gas. In this model, gas remains even at 10 billion years.

On the other hand, there is no gas in the Local Group dwarf spheroidal galaxies. This is because we compute the evolution of dwarf spheroidal galaxies with isolated models. The Local Group dwarf spheroidal galaxies lost gas by the ram pressure and tidal stripping [31]. These effects are not taken into account in these simulations. Since the aim of this study is to understand the enrichment of heavy elements at

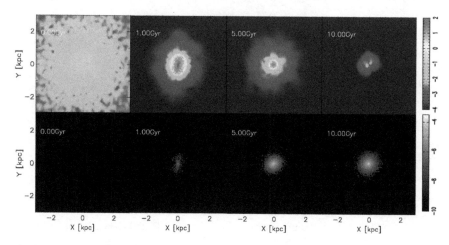

Fig. 3.2 Evolution of model s000 (Fig. 1 of [15] reproduced by permission of the AAS). Upper panels: slice gas density at 0, 1, 5, and 10 billion years in log scale from 10^{-4} cm^{-3} (blue) and 10^2 cm^{-3} (red). Lower panels: stellar surface density at 0, 1, 5, and 10 billion years with log scale from 10^0 $M_\odot$kpc^{-2} (black) and $10^{6.5}$ $M_\odot$kpc^{-2} (white)

early phases of galaxy evolution, the main results and conclusions are not affected by the lack of ram-pressure or tidal stripping.

The lower panel of Fig. 3.2 shows the time evolution of stellar surface density. According to these panels, we can see that stars are formed at the center of the galaxy and the galaxy expands from center to outer region. Size of the galaxy is ∼1 kpc at 10 billion years. This size is consistent with the size of Local Group dwarf spheroidal galaxies.

Figure 3.3 depicts time evolution of radial profiles. Figure 3.3a shows radial dark matter profiles. Central density increases due to the collapse of dark matter particles because we assume dark matters are not in dynamical equilibrium. The profile does not change over 10 billion years after the initial collapse.

Radial gas and stellar density profiles are related to each other. Figure 3.3b represents radial gas density profiles. As shown in this figure, the evolution of gas density follows that of dark matter. Gas density decreases from 1 to 10 billion years due to star formation and outflows induced by supernova feedback. Figure 3.3c denotes time evolution of radial stellar density. Stellar density increases as the star formation proceeds. The peak of the star formation is ∼2 billion years in this model. Due to the active star formation in this period, central stellar density increases ∼2 dex from 1 to 5 billion years. Stellar density profile is truncated at 1 kpc from the center of the galaxy. As we have stated before, this size is comparable to the size of Local Group dwarf spheroidal galaxies.

Figure 3.3d shows the stellar velocity dispersion profile. Velocity dispersion (σ_v) in a galaxy is mainly determined by the total mass of the galaxy. The average value of σ_v is ∼10 kms^{-1}. This value is consistent with the velocity dispersion of the Fornax dwarf spheroidal galaxy [48].

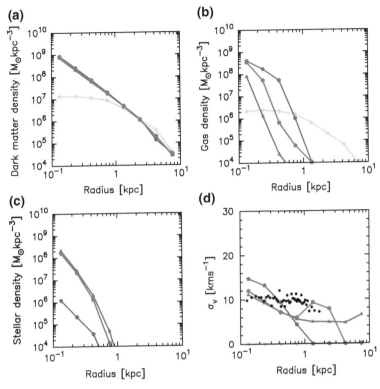

Fig. 3.3 Evolution of radial profiles in model s000 at 0 (green), 1 (blue), 5 (magenta), and 10 billion years (red) from the beginning of the simulation (Fig. 2 of [15] reproduced by permission of the AAS). Panels from **a** to **d** represent the evolution of the radial dark matter density profile, the radial gas density profile, the radial stellar density profile, and the radial stellar velocity dispersion profile. Black dots represent observed stellar velocity dispersion in the dwarf spheroidal galaxy Fornax [48]

Parameters of the simulations affect the evolution of radial profiles. Figure 3.4 shows parameter dependence on the radial profiles. Here we show the effects of threshold density for star formation (n_{th}), dimensionless star formation efficiency ($c_\star$), and the thermal energy of supernovae (ε_{SN}). Since these parameters are for baryon physics, dark matter profiles do not depend on them. Also, velocity dispersion profiles are not significantly affected by these parameters because they are determined by the total mass of the system, where the mass of dark matter is dominated. The value of ε_{SN} has the most significant impacts on the radial profiles, i.e., heating of interstellar medium by supernovae significantly affects the evolution of galaxies. In model se01, we reduce the value of ε_{SN} for 1 dex than model s000. Due to the insufficient feedback, stars are more easily formed than the other models. In this model, most gases have already consumed for star formation at 5 billion years (Fig. 3.4b, c). According to Fig. 3.4, the value of n_{th} also affects the radial profiles. Reducing this parameter makes star formation easier, i.e., n_{th} has similar effects on

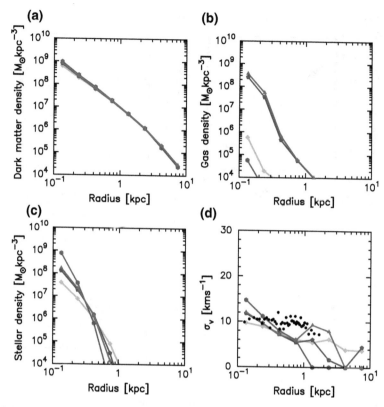

Fig. 3.4 Radial profiles (**a** dark matter density, **b** gas density, **c** stellar density, and **d** velocity dispersion) of models of different parameters at 5 billion years (Fig. 14 of [15], reproduced by permission of the AAS). Red triangles, blue pentagons, green diamonds, and magenta hexagons represent model s000, sc50 ($c_\star = 0.5$), sn10 ($n_{th} = 10\,cm^{-3}$), and se01 ($\varepsilon_{SN} = 10^{50}$ erg), respectively

ε_{SN}. On the other hand, the value of $c_\star$ does not have significant effects on radial profiles when n_{th} is set to be 100 cm^{-3}. This result is consistent with Saitoh et al. [39], which shows that the value of $c_\star$ does not significantly affect the evolutionary histories of galaxies if we set enough higher value of n_{th}.

3.3   Star Formation Histories

In this section, star formation histories of all models discussed in this book are presented. Figure 3.5 denotes the star formation rates as a function of time in model G. Star formation in this model starts when gas falls onto the center and meets the conditions of star formation. Star formation rates oscillate because supernova feedback randomly heats the gases in the interstellar medium. Star formation rates

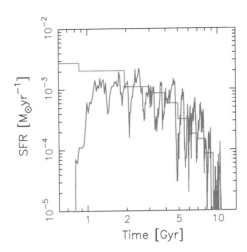

Fig. 3.5 Star formation rates (SFR) as functions of time in model G (red curve) (Fig. 4 of [17] reproduced under the terms of Creative Commons Attribution 3.0 license[5]). The green line represents the star formation history of the Sculptor dwarf spheroidal galaxy [8]. The unit of horizontal axis is billion years (Gyr). The SFRs are drawn with a unit of solar mass per year

in this model are $\sim 10^{-3}$ $M_\odot \mathrm{yr}^{-1}$, which are similar to those in Local Group dwarf spheroidal galaxies such as Fornax and Sculptor [7, 8].

The initial central density of halos significantly affects the time variations of star formation rates. Figure 3.6 represents the star formation histories computed in models with a different initial density of halos. Star formation rates in models A and B are $\lesssim 10^{-3}$ $M_\odot \mathrm{yr}^{-1}$. In these models, typical dynamical times ($t_{\mathrm{dyn}} = \sqrt{3/4\pi G \rho_{\mathrm{c}}}$, where ρ_{c} and G are the density of halos and the gravitational constant, respectively) are ~ 100 million years, i.e., gases initially collapse in this timescale. This timescale is longer than the lifetimes of progenitors of supernovae (~ 10 million years). In these models, heating of supernova feedback efficiently suppresses the formation of the next generation stars.

On the other hand, models C and D have the strong peak of star formation rates of $\gtrsim 10^{-2}$ $M_\odot \mathrm{yr}^{-1}$ at $\lesssim 100$ million years from the beginning of the simulation. In these models, typical dynamical times are comparable to the lifetimes of massive stars. Due to the short dynamical timescale, stars form before supernova feedback heats the gas in the star-forming region.

The total mass of halo is also an important parameter which affects the star formation histories. Figure 3.7 denotes star formation rates as functions of time in models with a different total mass of halos. Model F reaches the highest star formation rates because this model is heaviest among all models. Star formation rates in all models at $\lesssim 1$ billion years from the beginning of the simulation are suppressed due to heating of supernova feedback.

As is inferred from the radial profiles, the values of $\varepsilon_{\mathrm{SN}}$ and n_{th} have significant impacts on star formation histories. Figure 3.8 shows star formation histories computed with different baryonic parameters. Model se01 reduces the value of thermal feedback from supernovae. Stars are formed in cold gas clouds. Heating by supernova prevents star formation. As shown in this figure, reducing the value of the supernova feedback energy affects enhancing star formation rates. In the case of model sn10,

Fig. 3.6 Star formation rates as functions of time in models A (red curve), B (green curve), C (blue curve), and D (orange curve, Fig. 8 of [16] reproduced by permission of the Oxford University Press). Vertical dotted lines represent the time one billion years after the formation of first stars

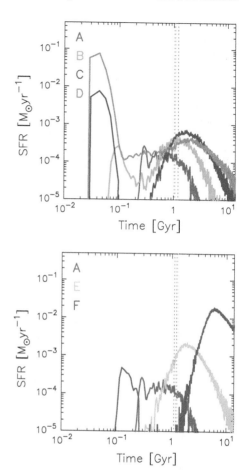

Fig. 3.7 Same as Fig. 3.6, but for models A (red), E (cyan), and F (magenta, Fig. 12 of [16] reproduced by permission of the Oxford University Press)

the lower value of the threshold density for star formation (n_{th}) shifts the peak of star formation in the earlier phase than that of model s000. This is because stars can be formed in more diffuse gas in this model.

On the other hand, model sc50 has similar star formation histories with model s000. This result means that the value of $c_\star$ does not affect the star formation histories when we adopt the sufficiently high value of n_{th}. This result is consistent with Saitoh et al. [39].

As shown in Fig. 3.8, the energy of supernova feedback and threshold density for star formation significantly affect the star formation histories in isolated dwarf galaxy models. Figure 3.9 compares star formation histories computed on models s000 and mExt. Model mExt adopts lower values of ε_{SN} and n_{th}. As is expected, star formation rates in model mExt get higher and the peak shift earlier phase compared to model s000. This difference would have significant impacts on the chemical evolution of galaxies (see Sect. 6.3).

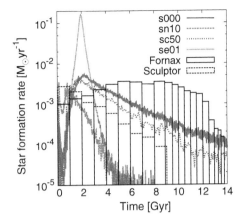

Fig. 3.8 Time variations of star formation rates for our models with different parameters (red curve: model s000, magenta dotted curve: model se01, green dashed curve: model se01, and blue short-dashed curve: model sc50, Fig. 15 of [15] reproduced by permission of the AAS). The black histogram and black-dotted histogram denote star formation histories of the dwarf spheroidal galaxies Fornax [7] and Sculptor [8], respectively

Fig. 3.9 Same as Fig. 3.8 but for models s000 (red-solid curve) and mExt (magenta-dashed curve, Fig. 3 of [15] reproduced by permission of the AAS). Blue and green-dashed histograms denote the star formation histories estimated from color magnitude diagrams of the Fornax [7] and the Sculptor dwarf spheroidal galaxies [8], respectively

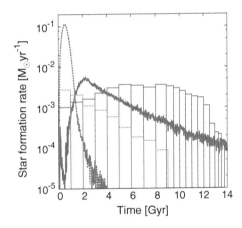

3.4 Dynamical Friction

Here we show the effects of dynamical friction. In this book, we assume the same number of gas and dark matter particles. To be consistent with the baryon fraction, $f_b = 0.15$ [35], we set $m_{DM} = 5.7\, m_{gas}$. Massive (dark matter) particles fall into the center by the dynamical friction. Provided that the density distribution is a singular isothermal sphere,

$$\rho(r) = \frac{v_c^2}{4\pi G r^2},\tag{3.4}$$

where v_c^2 is the constant circular speed, r is the radius, massive particles (dark matter particles) lose angular momentum (L) at a rate,

$$\frac{dL}{dt} = -0.428 \ln \Lambda \frac{G m_{DM}^2}{r}, \tag{3.5}$$

where

$$\Lambda = \frac{b_{max}}{b_{90}} \approx \frac{b_{max} v_t^2}{G m_{DM}}, \tag{3.6}$$

and b_{max}, b_{90}, v_t are the maximum value of impact parameter, 90° deflection radius, and typical relative velocity [3]. The angular momentum at the radius r in the singular isothermal sphere is $L = m_{DM} r v_c$. Substituting this to the Eq. (3.5),

$$r \frac{dr}{dt} = -0.302 \ln \Lambda \frac{G m_{DM}^2}{\sigma}. \tag{3.7}$$

Here we use $v_c = \sqrt{2}\sigma$. From the Eq. (3.7), we obtain the timescale that massive particles reach the center of the halo,

$$t_{fric} = \frac{19 \text{ billion years}}{\ln \Lambda} \left(\frac{r}{5 \text{ kpc}} \right)^2 \frac{\sigma}{200 \text{ kms}^{-1}} \frac{10^8 M_\odot}{m_{DM}}. \tag{3.8}$$

In the isolated dwarf galaxy model adopted in this book, typical values are $b_{max} = 1$ kpc, $m_{DM} = 10^4 M_\odot$, and $v_t \approx \sigma = 10$ kms^{-1}. In this case, $\ln \Lambda = 7.75$. From Eq. (3.8), the timescale of dynamical friction is $t_{fric} = 49.0$ billion years. This timescale is sufficiently longer than the cosmic time. On the other hand, in the case of simulations of massive galaxies ($b_{max} = 5$ kpc, $v_t \approx \sigma = 200$ kms^{-1}) with $m_{DM} = 10^8 M_\odot$ (e.g., [22, 27]), $t_{fric} = 3.1$ billion years. In this case, dynamical friction affects the evolution of simulated galaxies.

To confirm that the dynamical friction does not affect the results of this study, we performed a series of simulations adopting $m_{DM} = m_{gas}$. Other parameters are the same as those in model G. Figure 3.10 shows radial density profiles of dark matter, gas, and stars. According to this figure, radial profiles of both models G and Q are overlapped. Figure 3.11 denotes the time variations of star formation rates in these models. Both models show star formation rates of $\sim 10^{-3} M_\odot \text{yr}^{-1}$. We cannot see any significant difference between models G and Q. These results indicate that dynamical friction due to adopting particles with different mass does not affect the results of this study.

Fig. 3.10 Radial density profiles of dark matter (black), gas (red) and stars (blue) at 13.8 billion years from the beginning of the simulation. Solid and dashed curves denote the models G and Q. Both curves are mostly overlapped

Fig. 3.11 Star formation rates as functions of time in models G (blue) and Q (red)

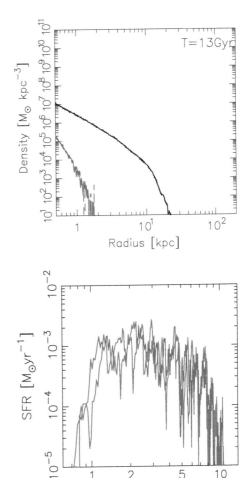

3.5 Metallicity Distribution

Metallicity distribution is an important indicator to discuss the chemical evolution of galaxies. Figure 3.12 shows the metallicity distribution of model s000, Fornax, and Sculptor dwarf spheroidal galaxies. Fe is produced by both core-collapse supernovae and type Ia supernovae. Since model s000 does not include the effect of type Ia supernovae, the peak of metallicity in model s000 is lower than those of Fornax and Sculptor dwarf spheroidal galaxies.

When we include the effects of type Ia supernovae, computed metallicity distribution matches with observed dwarf spheroidal galaxies with similar mass. Figure 3.13 represents the metallicity distribution computed in model G at 13.8 billion years from the start of the simulation. The median metallicity at 13.8 billion years is

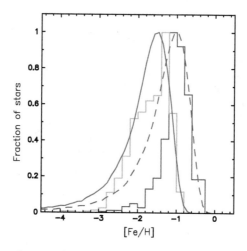

Fig. 3.12 Comparison of computed and observed metallicity distribution functions. Red curve and magenta-dashed curve represent metallicity distribution functions computed in model s000 and shifted metallicity distribution for 0.5 dex in model s000, respectively. Blue and green histogram show that the observed metallicity distributions of the Fornax dwarf spheroidal galaxy [25] and the Sculptor dwarf spheroidal galaxy [23–25], respectively (Fig. 4 of [15] reproduced by permission of the AAS)

Fig. 3.13 Metallicity distribution functions in model G at 13.8 billion years from the beginning of the simulation (red histogram), observed dwarf spheroidal galaxies Sculptor (green dashed histogram), Leo I (blue dash-dotted histogram, Fig. 5 of [17] reproduced under the terms of Creative Commons Attribution 3.0 license[5]). Observed values are taken from [23–25, 45]

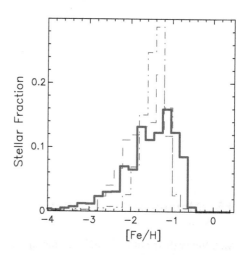

[Fe/H] $= -1.34$. The final stellar mass is $M_\star = 3.72 \times 10^6 \, M_\odot$. These results are consistent with the dwarf spheroidal galaxies Sculptor ([Fe/H] $= -1.68$ and $M_\star = 3.9 \times 10^6 \, M_\odot$) and Leo I ([Fe/H] $= -1.45$ and $M_\star = 4.9 \times 10^6 \, M_\odot$) [26].

   Metallicity distribution functions reflect the star formation histories. Figure 3.14 shows the computed metallicity distribution functions in models A, B, C, D (panel a) and models A, E, F (panel b). Models with higher star formation rates show steeper slopes of metallicity distribution functions at low metallicity (Figs. 3.6 and 3.14a).

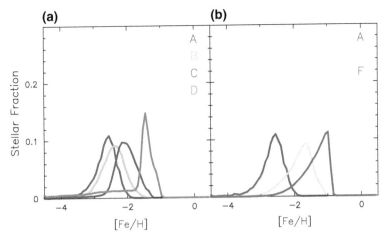

Fig. 3.14 Stellar metallicity distribution functions at 14 billion years (Fig. 2 of [16] reproduced by permission of the Oxford University Press). **a** Metallicity distribution functions of models A (red), B (green), C (blue), and D (orange). **b** Same as **a**, but for models A (red), E (cyan), and F (magenta)

Table 3.2 Metallicity and stellar mass at 14 billion years from the beginning of the simulation

Model	$\langle$[Fe/H]$\rangle$	σ	$M_\star (10^6 \, M_\odot)$
A	−2.6	0.35	0.3
B	−2.4	0.39	1.2
C	−2.0	0.40	3.4
D	−1.5	0.60	6.4
E	−1.8	0.44	8.0
F	−1.3	0.42	170.0

The columns from left to right represent name of models, the median stellar metallicity ($\langle$[Fe/H]$\rangle$), the standard deviation of metallicity distribution function (σ), and stellar mass ($M_\star$)

The sharp cut-off seen in the metallicity distribution function in model F reflects that this model still forms stars at the end of the simulation (Figs. 3.7 and 3.14b). Table 3.2 lists the final properties of models A, B, C, D, E, and F.

Figure 3.15 shows gas metallicity distribution functions at 0.2 billion years from the beginning of the star formation. According to this figure, models A, B, E, and F have larger scatters in gas phase metallicity distribution functions than those in models C and D. Section 5.3 shows that this inhomogeneity in metallicity affects the scatters of [Eu/Fe].

Metallicity distribution functions are also affected by parameters adopted in simulations. Figure 3.16 shows metallicity distribution functions with different parameters of baryon physics. Model se01 reaches the highest metallicity among models compared in this figure because the peak of the star formation rates is the highest among the compared models (Fig. 3.8). Metallicity in model sn10 is lower than the other models. This is because star formation stops at earlier phases in this model.

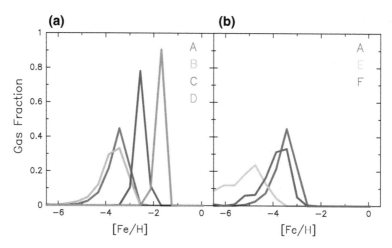

Fig. 3.15 Same as Fig. 3.14, but for gas phase metallicity distribution functions at 0.2 billion years from the beginning of the star formation (Fig. 3 of [16] reproduced by permission of the Oxford University Press)

Fig. 3.16 Same as Fig. 3.12, but for models s000 (red curve), sn10 (green dashed curve), sc50 (blue dashed-dotted curve), and se01 (magenta dashed-dotted-dotted curve, Fig. 16 of [15] reproduced by permission of the AAS). The black and black-dotted histograms are the metallicity distribution of the Fornax dwarf spheroidal galaxy [25] and the Sculptor dwarf spheroidal galaxy [23–25], respectively

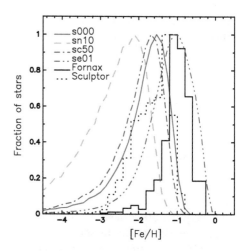

Model sc50 has similar metallicity distribution due to the similar star formation histories with model s000.

3.6   Mass-Metallicity Relation

Figure 3.17 compares the computed and observed mass-metallicity relation in dwarf galaxies. As shown in this figure, both models and observations show the increasing trend toward heavier galaxies. Models with higher central densities and total masses

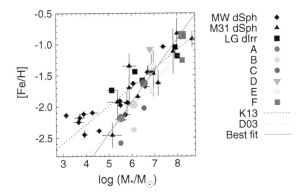

Fig. 3.17 Computed and observed stellar mass-metallicity relations (Fig. 4 of [16] reproduced by permission of the Oxford University Press). Red, green, blue filled circles, orange inverted triangle, cyan, and magenta filled squares represent models A, B, C, D, E, and F, respectively. Original values of these models are plotted with small plots. Expected values if we include the effects of type Ia supernovae are plotted large plots. The observed values of the Local Group dwarf irregular galaxies, M31 dwarf spheroidal galaxies, and the Milky Way dwarf spheroidal galaxies are shown with black squares, triangles, and diamonds, respectively ([26], K13). The purple dashed, sky-blue dot-dashed, and green solid lines respectively denote the results of least-square fitting with the sample of K13: $\langle[\text{Fe/H}]\rangle = (-1.69 \pm 0.04) + (0.30 \pm 0.02)\log\left(M_*/10^6 M_\odot\right)$, [10]: $[\text{Fe/H}] \propto M_*^{0.40}$ (D03), and models except for model D: $\langle[\text{Fe/H}]\rangle = (-1.93 \pm 0.05) + (0.50 \pm 0.04)\log\left(M_*/10^6 M_\odot\right)$

have higher median metallicity and stellar mass. Gas and metals are more difficult to be removed from heavier or higher central density models due to deeper gravitational potential wells (e.g., [9]). In the context of mass-metallicity relation, the central density and total mass of halos show similar effects to the metallicity and stellar mass.

The models show lower median metallicity than that of the observations because we ignore the effects of type Ia supernovae (small colored plots in Fig. 3.17). The metallicity becomes consistent with the observations if we shift the metallicity with 0.4 dex (large colored plots in Fig. 3.17). The contribution of type Ia supernovae to the total amount of Fe is estimated to be ~60–65% by using the observations of [α/Fe] ratios as a function of [Fe/H] in the Milky Way halo [13, 36]. The degree of the contribution of type Ia supernovae differs among galaxies with different star formation histories.

Model D deviates from the mass-metallicity relation. Model D efficiently retains metals in a galaxy due to its very high-density halo. As shown in Fig. 3.1, the assumption of the cut off the dark matter density profile is unnatural. We can exclude this model from the model of Local Group dwarf galaxies in the context of the mass-metallicity relation.

The predicted mass-metallicity relation has a steeper slope than the observed one. Several simulations also reported that steeper slopes of mass-metallicity relation than the observed relation (e.g., [6, 30, 41, 46]). Tidal disruptions, which are not included in the models, may produce shallower slopes. These effects can reduce the stellar masses of smaller galaxies (e.g., [31]). Treatment of stellar feedback may also affect the mass-metallicity relation. In this simulation, feedback from supernovae is implemented in the thermal form. Okamoto et al. [34] pointed out that the radiation-pressure from young massive stars efficiently suppresses the star formation at high metallicity. They showed that this effect makes the slope shallower.

3.7 Enrichment of α-Elements

The metallicity where α-elements to Fe ratios (e.g., [Mg/Fe]) begins to decrease can be an indicator for the speed of chemical evolution. Delayed production of Fe from type Ia supernovae, which synthesize few α-elements, reduces the ratios of [Mg/Fe]. Figure 3.18 depicts [Mg/Fe] as a function of [Fe/H] in model G. The ratios of [Mg/Fe] start to decrease at [Fe/H] $\gtrsim -2.5$ in this model. This metallicity is consistent with that of the Sculptor dwarf spheroidal galaxy [45]. This model shows scatters of [Mg/Fe] ratios less than 0.1 dex at [Fe/H] $\lesssim -2.8$. This result is also consistent with the observations in the Local Group galaxies (e.g., [11]).

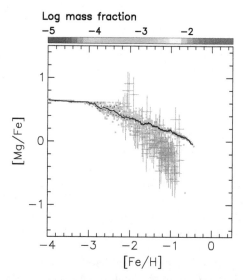

Fig. 3.18 [Mg/Fe] as functions of [Fe/H] computed in model G at 13.8 billion years from the beginning of the simulation (Fig. 6 of [17] reproduced under the terms of Creative Commons Attribution 3.0 license (https://creativecommons.org/licenses/by/3.0/)). The mass fraction of stellar mass is plotted with color-code in the logarithmic scale. The median and scatters of computed [Mg/Fe] ratios are shown with black solid and dotted curves, respectively. Gray dots represent the observed [Mg/Fe] ratios in the dwarf spheroidal galaxy Sculptor [25]. Errors of [Fe/H] and [Mg/Fe] in the plotted data are restricted to be Δ[Fe/H] < 0.15 and Δ[Mg/Fe] < 0.30 to prevent scatters due to observational errors [14]

The computed ratios of [Mg/Fe] decreases more slowly toward higher metallicity than the observed ones. This difference may be caused by the difference of the star formation histories between model G and the observations. The Sculptor dwarf spheroidal galaxy has decreasing rates of star formation [8]. On the other hand, model G continuously forms stars over 9 billion years (see Fig. 3.5). Homma et al. [18] showed that the steeper slope of [Mg/Fe] ratios can be seen in the models which have the peak of the star formation rates at the earlier phases.

Delay time distributions of type Ia supernovae may also affect the slope of [Mg/Fe] ratios. In model G, we assume the power-law delay time distribution with the minimum delay time of 0.1 billion years. Kobayashi and Nomoto [28] showed that the sufficiently low rate of type Ia supernova is required to reproduce the observed ratios of α-elements to Fe. Homma et al. [18] favored the minimum delay time of 0.5 billion years to reproduce the slope of [Mg/Fe] ratios in dwarf spheroidal galaxies.

Figure 3.19 compares [Mg/Fe] as a function of [Fe/H] for models with different central densities and the total mass of halos at one billion years from the beginning of the simulation. All models plotted in this figure do not assume the effects of type Ia supernovae. The ratios of [Mg/Fe] are, therefore, almost constant because

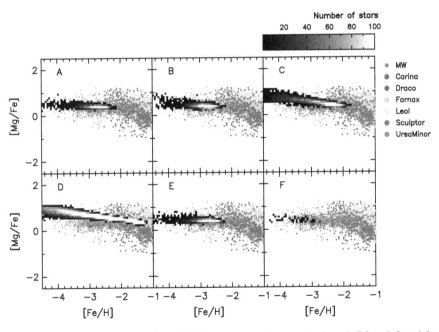

Fig. 3.19 [Mg/Fe] ratios as functions of [Fe/H] computed in models A, B, and, C from left to right in the top panel and models D, E, and F from left to right in the bottom panel. All models are plotted at one billion years from the beginning of the star formation (Fig. 5 of [16] reproduced by permission of the Oxford University Press). Gray scales denote the number of stars. The plots represent the observed value of the Milky Way (sky-blue), Carina (red), Draco (blue), Fornax (green), Leo I (cyan), Sculptor (magenta), and Ursa Minor (orange) dwarf spheroidal galaxies (SAGA database [42–44, 49])

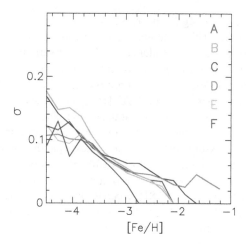

Fig. 3.20 Standard deviations of [Mg/Fe] ratios (σ) as functions of [Fe/H] in models A (red), B (green), C (blue), D (orange), E (cyan), and F (magenta) at one billion years from the first star formation (Fig. 6 of [16] reproduced by permission of the Oxford University Press)

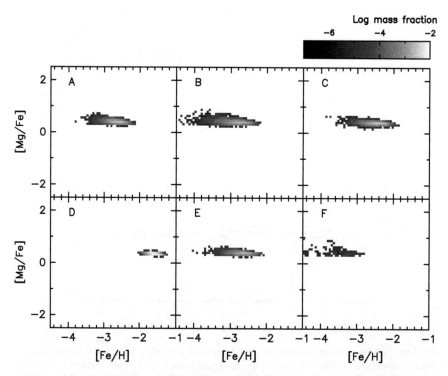

Fig. 3.21 Gas phase [Mg/Fe] ratios as functions of [Fe/H] at one billion years from the beginning of the simulation in models A, B, and C from left to right in the top panel and models D, E, and F from left to right in the bottom panel (Fig. 7 of [16] reproduced by permission of the Oxford University Press). Mass fraction is coded with color from black to white in logarithmic scale

core-collapse supernovae produce both Mg and Fe. Slightly decreasing trend of [Mg/Fe] ratios is due to the effects of variations of yields in core-collapse supernovae from different mass progenitors. The effects of early star formation efficiencies are clearly seen in models A, B, C, and D. Models with higher star formation efficiencies such as models C and D reaches higher metallicities than those of models A and B. The effects of total mass of halo cannot be clearly seen in [Mg/Fe] ratios.

All models exhibit small scatters of [Mg/Fe] ratios. Figure 3.20 shows scatters of [Mg/Fe] as a function of [Fe/H]. As shown in this figure, standard deviations of [Mg/Fe] in all models are lower than 0.2. The standard deviation decreases toward higher metallicity. This result means that the spatial variations of metallicity become more homogeneous at higher metallicity.

Figure 3.21 denotes gas phase [Mg/Fe] as a function of [Fe/H] at one billion years from the beginning of the simulation. As shown in this figure, there are no significant scatters of gas phase [Mg/Fe] ratios. From these results, we can confirm that the metal mixing models properly work to erase unnatural scatters.

3.8 Summary

In this chapter, we describe the chemodynamical evolution of the isolated dwarf galaxy models. We compare the results of radial profiles, star formation histories, metallicity distribution functions, mass-metallicity relation, and α-element abundances to the observed Local Group dwarf galaxies. We confirm that these properties are consistent with observations.

References

1. Baba J, Saitoh TR, Wada K (2013) Dynamics of nonsteady spiral arms in disk galaxies. Astrophys J 763(46):46
2. Begeman KG, Broeils AH, Sanders RH (1991) Extended rotation curves of spiral galaxies—Dark haloes and modified dynamics. Mon Not R Astron Soc 249:523–537
3. Binney J, Tremaine S (2008) Galactic dynamics, 2nd edn. Princeton University Press
4. Carraro G, Chiosi C, Girardi L, Lia C (2001) Dwarf elliptical galaxies: structure, star formation and colour-magnitude diagrams. Mon Not R Astron Soc 327:69–79
5. Chiosi C, Carraro G (2002) Formation and evolution of elliptical galaxies. Mon Not R Astron Soc 335:335–357
6. Davé R, Katz N, Oppenheimer BD, Kollmeier JA, Weinberg DH (2013) The neutral hydrogen content of galaxies in cosmological hydrodynamic simulations. Mon Not R Astron Soc 434:2645–2663
7. de Boer TJL, Tolstoy E, Hill V, Saha A, Olszewski EW, Mateo M, Starkenburg E, Battaglia G, Walker MG (2012a) The star formation and chemical evolution history of the Fornax dwarf spheroidal galaxy. Astron Astrophys 544:A73
8. de Boer TJL, Tolstoy E, Hill V, Saha A, Olsen K, Starkenburg E, Lemasle B, Irwin MJ, Battaglia G (2012b) The star formation and chemical evolution history of the sculptor dwarf spheroidal galaxy. Astron Astrophys 539:A103

9. Dekel A, Silk J (1986) The origin of dwarf galaxies, cold dark matter, and biased galaxy formation. Astrophys J 303:39–55
10. Dekel A, Woo J (2003) Feedback and the fundamental line of low-luminosity low-surface-brightness/dwarf galaxies. Mon Not R Astron Soc 344:1131–1144
11. Frebel A, Norris JE (2015) Near-field cosmology with extremely metal-poor stars. Annu Rev Astron Astrophys 53:631–688
12. Fujii MS, Baba J, Saitoh TR, Makino J, Kokubo E, Wada K (2011) The dynamics of spiral arms in pure stellar disks. Astrophys J 730(109):109
13. Goswami A, Prantzos N (2000) Abundance evolution of intermediate mass elements (C to Zn) in the MilkyWay halo and disk. Astron Astrophys 359:191–212
14. Hill V, DART Collaboration (2012) Abundance patterns and the chemical enrichment of nearby dwarf galaxies. In: Aoki W, Ishigaki M, Suda T, Tsujimoto T, Arimoto N (eds) Galactic archaeology: near-field cosmology and the formation of the milky way, vol 458, Astronomical society of the pacific conference series, p 297
15. Hirai Y, Ishimaru Y, Saitoh TR, Fujii MS, Hidaka J, Kajino T (2015) Enrichment of r-process elements in dwarf spheroidal galaxies in chemodynamical evolution model. Astrophys J 814:41
16. Hirai Y, Ishimaru Y, Saitoh TR, Fujii MS, Hidaka J, Kajino T (2017) Early chemo-dynamical evolution of dwarf galaxies deduced from enrichment of r-process elements. Mon Not R Astron Soc 466:2474–2487
17. Hirai Y, Saitoh TR, Ishimaru Y, Wanajo S (2018) Enrichment of Zinc in galactic chemodynamical evolution models. Astrophys J 855(63):63
18. Homma H, Murayama T, Kobayashi MAR, Taniguchi Y (2015) A new chemical evolution model for dwarf spheroidal galaxies based on observed long star formation histories. Astrophys J 799:230
19. Hopkins PF, Quataert E, Murray N (2011) Self-regulated star formation in galaxies via momentum input from massive stars. Mon Not R Astron Soc 417:950–973
20. Hopkins PF, Kereš D, Oñorbe J, Faucher-Giguère CA, Quataert E, Murray N, Bullock JS (2014) Galaxies on FIRE (Feedback In Realistic Environments): stellar feedback explains cosmologically inefficient star formation. Mon Not R Astron Soc 445:581–603
21. Hu CY, Naab T, Glover SC, Walch S, Clark PC (2017) Variable interstellar radiation fields in simulated dwarf galaxies: supernovae versus photoelectric heating. Mon Not R Astron Soc 471:2151–2173
22. Katz N, Gunn JE (1991) Dissipational galaxy formation. I—Effects of gasdynamics. Astrophys J 377:365–381
23. Kirby EN, Cohen JG (2012) Detailed abundances of two very metal-poor stars in dwarf galaxies. Astron J 144:168
24. Kirby EN, Guhathakurta P, Bolte M, Sneden C, Geha MC (2009) Multielement abundance measurements from medium-resolution spectra. I. The sculptor dwarf spheroidal galaxy. Astrophys J 705:328–346
25. Kirby EN, Guhathakurta P, Simon JD, Geha MC, Rockosi CM, Sneden C, Cohen JG, Sohn ST, Majewski SR, Siegel M (2010) Multi-element abundance measurements from medium-resolution spectra. II. Catalog of stars in Milky Way dwarf satellite galaxies. Astrophys J Suppl Ser 191:352–375
26. Kirby EN, Cohen JG, Guhathakurta P, Cheng L, Bullock JS, Gallazzi A (2013) The universal stellar mass-stellar metallicity relation for dwarf galaxies. Astrophys J 779:102
27. Kobayashi C (2004) GRAPE-SPH chemodynamical simulation of elliptical galaxies—I. Evolution of metallicity gradients. Mon Not R Astron Soc 347:740–758
28. Kobayashi C, Nomoto K (2009) The role of type Ia supernovae in chemical evolution. I. Lifetime of type Ia supernovae and metallicity effect. Astrophys J 707:1466–1484
29. Kumamoto J, Baba J, Saitoh TR (2017) Imprints of zeroage velocity dispersions and dynamical heating on the age-velocity dispersion relation. Publ Astron Soc Jpn 69(32):32
30. Lu Y, Wechsler RH, Somerville RS, Croton D, Porter L, Primack J, Behroozi PS, Ferguson HC, Koo DC, Guo Y, Safarzadeh M, Finlator K, Castellano M, White CE, Sommariva V, Moody C (2014) Semianalytic models for the CANDELS survey: comparison of predictions for intrinsic galaxy properties. Astrophys J 795:123

31. Nichols M, Revaz Y, Jablonka P (2014) Gravitational tides and dwarf spheroidal galaxies. Astron Astrophys 564:A112

32. Oh S-H, de Blok WJG, Brinks E, Walter F, Kennicutt RC Jr (2011) Dark and luminous matter in THINGS dwarf galaxies. Astron J 141:193

33. Oh S-H, Hunter DA, Brinks E, Elmegreen BG, Schruba A, Walter F, Rupen MP, Young LM, Simpson CE, Johnson MC, Herrmann KA, Ficut- Vicas D, Cigan P, Heesen V, Ashley T, Zhang H-X (2015) High-resolution mass models of dwarf galaxies from LITTLE THINGS. Astron J 149:180

34. Okamoto T, Shimizu I, Yoshida N (2014) Reproducing cosmic evolution of galaxy population from z = 4 to 0. Publ Astron Soc Jpn 66:70

35. Planck Collaboration (2014) Planck 2013 results. XVI. Cosmological parameters. Astron Astrophys 571:A16

36. Prantzos N (2008) The metallicity distribution of the halo and the satellites of the Milky Way in the hierarchical merging paradigm. Astron Astrophys 489:525–532

37. Revaz Y, Jablonka P (2012) The dynamical and chemical evolution of dwarf spheroidal galaxies with GEAR. Astron Astrophys 538:A82

38. Revaz Y, Jablonka P, Sawala T, Hill V, Letarte B, Irwin M, Battaglia G, Helmi A, Shetrone MD, Tolstoy E, Venn KA (2009) The dynamical and chemical evolution of dwarf spheroidal galaxies. Astron Astrophy 501:189–206

39. Saitoh TR, Daisaka H, Kokubo E, Makino J, Okamoto T, Tomisaka K, Wada K, Yoshida N (2008) Toward first-principle simulations of galaxy formation: I. How should we choose star-formation criteria in high-resolution simulations of disk galaxies? Publ Astron Soc Jpn 60:667–681

40. Schaye J, Crain RA, Bower RG, Furlong M, Schaller M, Theuns T, Dalla Vecchia C, Frenk CS, McCarthy IG, Helly JC, Jenkins A, Rosas- Guevara YM, White SDM, Baes M, Booth CM, Camps P, Navarro JF, Qu Y, Rahmati A, Sawala T, Thomas PA, Trayford J (2015) The EAGLE project: simulating the evolution and assembly of galaxies and their environments. Mon Not R Astron Soc 446:521–554

41. Somerville RS, Davé R (2015) Physical models of galaxy formation in a cosmological framework. Annu Rev Astron Astrophys 53:51–113

42. Suda T, Katsuta Y, Yamada S, Suwa T, Ishizuka C, Komiya Y, Sorai K, Aikawa M, Fujimoto MY (2008) Stellar abundances for the galactic archeology (SAGA) database—compilation of the characteristics of known extremely metal-poor stars. Publ Astron Soc Jpn 60:1159–1171

43. Suda T, Yamada S, Katsuta Y, Komiya Y, Ishizuka C, Aoki W, Fujimoto MY (2011) The stellar abundances for galactic archaeology (SAGA) data base—II. Implications for mixing and nucleosynthesis in extremely metal-poor stars and chemical enrichment of the galaxy. Mon Not R Astron Soc 412:843–874

44. Suda T, Hidaka J, Ishigaki M, Katsuta Y, Yamada S, Komiya Y, Fujimoto MY, Aoki W (2014) Stellar abundances for galactic archaeology database for stars in dwarf galaxies. memsai 85, p 600

45. Suda T, Hidaka J, Aoki W, Katsuta Y, Yamada S, Fujimoto MY, Ohtani Y, Masuyama M, Noda K, Wada K (2017) Stellar abundances for galactic archaeology database. IV. Compilation of stars in dwarf galaxies. Publ Astron Soc Jpn 69:76

46. Torrey P, Vogelsberger M, Genel S, Sijacki D, Springel V, Hernquist L (2014) A model for cosmological simulations of galaxy formation physics: multi-epoch validation. Mon Not R Astron Soc 438:1985–2004

47. Valcke S, de Rijcke S, Dejonghe H (2008) Simulations of the formation and evolution of isolated dwarf galaxies. Mon Not R Astron Soc 389:1111–1126

48. Walker MG, Mateo M, Olszewski EW, Peñarrubia J, Wyn Evans N, Gilmore G (2009) A universal mass profile for dwarf spheroidal galaxies? Astrophys J 704:1274–1287

49. Yamada S, Suda T, Komiya Y, Aoki W, Fujimoto MY (2013) The stellar abundances for galactic archaeology (SAGA) database—III. Analysis of enrichment histories for elements and two modes of star formation during the early evolution of the MilkyWay. Mon Not R Astron Soc 436:1362–1380

Chapter 4
Enrichment of Zinc in Dwarf Galaxies

Abstract This chapter aims to clarify the role of supernovae (core-collapse supernovae, electron-capture supernovae, hypernovae, and type Ia supernovae) to the chemical evolution of galaxies. We focus on the enrichment of Zn in this chapter. Zn is the heaviest iron-peak element and has been used as a tracer of metallicity in damped Lyman-α systems. High-resolution spectroscopic observations have shown that there is a decreasing trend of [Zn/Fe] toward higher metallicity. However, the astrophysical sites and the enrichment of Zn are not well understood. In this chapter, we show that electron-capture supernovae can contribute to the enrichment of Zn at low metallicity. We find that stars which have [Zn/Fe] $\gtrsim 0.5$ are formed from the ejecta of electron-capture supernovae. At higher metallicity, type Ia supernovae make scatters of [Zn/Fe] ratios. We also find that it is difficult to reproduce observations of [Zn/Fe] ratios without considering the effects of electron-capture supernovae or low mass iron core-collapse supernovae. These results suggest that supernovae from the lowest mass ranges can be one of the main astrophysical sites of Zn [(Contents in this chapter have been published in Hirai et al. (Astrophys J 855:63, 2018 [9]) reproduced under the terms of Creative Commons Attribution 3.0 license (https://creativecommons.org/licenses/by/3.0/))].

4.1 Astrophysical Sites of Zn

The astrophysical sites of Zn are not well understood. Core-collapse supernovae can play an important role in the enrichment of Zn in galaxies. The iron core-collapse supernovae from stars with $\gtrsim 10\ M_\odot$ synthesize ^{64}Zn in the complete Si-burning [11, 36]. At higher metallicity, isotopes of $^{66-70}$Zn are synthesized by the neutron-capture process. The enrichment of Zn from iron core-collapse supernovae has been studied by several Galactic chemical evolution models [4, 6, 10–12, 29]. Timmes et al. [29] showed that it was necessary to reduce the yields of Fe by a factor of two to explain the [Zn/Fe] ratios. Kobayashi et al. [11] have shown that the yields of Nomoto et al. [15] could not produce enough Zn to be consistent with the observations.

© Springer Nature Singapore Pte Ltd. 2019
Y. Hirai, *Understanding the Enrichment of Heavy Elements
by the Chemodynamical Evolution Models of Dwarf Galaxies*,
Springer Theses, https://doi.org/10.1007/978-981-13-7884-3_4

Hypernovae have been regarded as one of the promising astrophysical sites [11, 30–32]. They produce 10 times larger kinetic energy than that of the normal iron core-collapse supernovae (e.g., [26]). Long gamma-ray bursts are thought to be associated with hypernovae [7, 17]. Their extended Si-burning regions synthesize a large amount of Zn. Compared to the yields of Nomoto et al. [15], ten times larger ratios of [Zn/Fe] could be seen if 50% of stars with $\geq 20M_\odot$ become hypernovae [11]. Tominaga et al. [30] showed that the trend of [Zn/Fe] ratios could be explained if yields of hypernovae with different progenitor masses are directly reflected the Zn abundances of extremely metal-poor stars.

Electron-capture or iron core-collapse supernovae from low mass progenitors can be another possible astrophysical site. Two-dimensional hydrodynamic simulations showed that electron-capture supernovae produced neutron-rich ejecta ($Y_e \sim 0.4 - 0.5$ [34]). Wanajo et al. [35] reported that electron-capture and low mass iron core-collapse supernovae could synthesize all stable isotopes of Zn. These supernovae cannot synthesize a significant amount of Fe, resulting in the yield of high [Zn/Fe] ratios. However, the enrichment of Zn from electron-capture supernovae in galaxies has not been studied.

The aim of this chapter is to clarify the roles of different types of supernovae on the enrichment of Zn. Section 4.2 shows the enrichment of Zn in our dwarf galaxy models. Section 4.3 presents the effects of mass ranges of electron-capture supernovae. Section 4.4 discusses the effects of electron-capture supernovae and hypernovae on the enrichment of Zn. Section 4.5 presents the effects of yields and the dependence on the resolution of the simulation. Section 4.6 summarizes the result of this chapter. In Table 4.1, we list the models discussed in this chapter.

Table 4.1 Models adopted in this chapter

Model	Number of particles	Mass range of electron-capture supernovae	f_{HN}	Supernova yields
G	2^{18}	Doherty et al. [3]	0.05	Nomoto et al. [14]
H	2^{18}	Poelarends [18]	0.05	Nomoto et al. [14]
I	2^{18}	8.5–9.0 $M_\odot$	0.05	Nomoto et al. [14]
J	2^{18}		0.5	Nomoto et al. [14]
K	2^{18}		0.05	Nomoto et al. [14]
L	2^{18}	Doherty et al. [3]		Nomoto et al. [14]
M	2^{18}	8.5–9.0 $M_\odot$		Nomoto et al. [14]
N	2^{18}	Doherty et al. [3]	0.05	Chieffi and Limongi [2]
O	2^{17}	Doherty et al. [3]	0.05	Nomoto et al. [14]
P	2^{16}	Doherty et al. [3]	0.05	Nomoto et al. [14]

Columns from left to right show the names of models, the total number of particles, the mass ranges of progenitors of electron-capture supernovae, the hypernova fractions (f_{HN}), and the adopted yields

4.2 Enrichment of Zn

4.2.1 Enrichment of Zn at Low Metallicity

In this subsection, we show the enrichment of Zn at [Fe/H] $\lesssim$ −2.5. Figure 4.1 denotes [Zn/Fe] as a function of [Fe/H] in model G. As shown in this figure, several stars are [Zn/Fe] $\gtrsim$ 0.5. We find that these stars reflect the ejecta from electron-capture supernovae. We also find that hypernovae increase the median values of [Zn/Fe] ratios. This value is related to the rates of hypernovae and electron-capture supernovae.

The ratios of [Zn/Fe] show a decreasing trend toward higher metallicity. At the lowest metallicity (−4.0 < [Fe/H] < −2.5), the chi-squared linear fitting shows that the slope of the [Zn/Fe] ratios is −0.12 ± 0.01. In this metallicity range, there are meager contributions of type Ia supernovae. The Milky Way halo has the steeper slope (−0.26 ± 0.04) than the computed one. The slope of [Zn/Fe] ratios may come from the mass ranges of electron-capture supernovae. The shallower slope in model G than the observation implies that the mass range of electron-capture supernovae is wider at lower metallicity than the stellar evolution model of Doherty et al. [3]. Stars

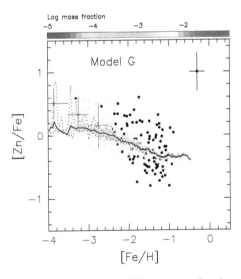

Fig. 4.1 Computed (model G) and observed [Zn/Fe] ratios as a function of [Fe/H]. We display the stellar mass fractions from 10^{-5} (blue) to 10^{-1} (red) in the logarithmic scale. The solid curves show the median [Fe/H] ratios in model G. The dotted curves represent the 5 and 95% data from the lowest value of [Zn/Fe]. The observed data of the Sculptor dwarf spheroidal galaxy are plotted with black dots [5, 21–24]. In the top right corner, the typical error bar of the observed [Zn/Fe] and [Fe/H] is plotted [24]. The red plots and vertical error bars respectively represent the mean and the maximum and minimum values of [Zn/Fe] ratios in the Milky Way halo binned in the range of horizontal error bars [20]

with high [Zn/Fe] ratios tend to form at lower metallicity. The highest ratio of [Zn/Fe] is 0.4 at [Fe/H] = −2.5. On the other hand, at [Fe/H] = −3.5, this value becomes 1.0. At lower metallicity, the low abundance of Fe and inhomogeneous metallicity distribution make it easy to form stars with high [Zn/Fe] ratios.

Figure 4.2 denotes the [Fe/H] (a) and [Zn/Fe] (b) ratios as functions of time. Figure 4.2a indicates that stars with [Fe/H] < −2 are produced within two billion years from the beginning of the simulation. All stars with [Zn/Fe] > 0.5 are also formed during this era (Fig. 4.2b). Spatial variations of Zn are significantly large at low metallicity due to the low rates of electron-capture supernovae. In model G, we assume 3.1% of all supernovae causes electron-capture supernovae at [Fe/H] = −3. The inhomogeneity of metals is erased due to a metal mixing at later phases of galaxy evolution.

Fig. 4.2 [Fe/H] **a** and [Zn/Fe] **b** as functions of time from the beginning of the simulation in model G (Fig. 8 of [9], reproduced under the terms of Creative Commons Attribution 3.0 license (https://creativecommons.org/licenses/by/3.0/)). The stellar mass fractions from 10^{-5} (blue) to 10^{-1} (red) are displayed in the logarithmic scale

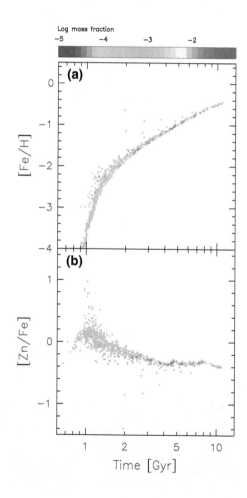

4.2.2 Enrichment of Zn at High Metallicity

Here we discuss the enrichment of Zn at [Fe/H] $\gtrsim$ −2.5. At −2.5 $\lesssim$ [Fe/H] $\lesssim$ −1.0, type Ia supernovae reduce [Zn/Fe] ratios. This is because type Ia supernovae do not produce a significant amount of Zn but synthesize a large amount of Fe. Stars with [Zn/Fe] < −0.5 are highly influenced by the ejecta from type Ia supernovae (Figs. 4.1 and 4.2b).

Observations of the Sculptor dwarf spheroidal galaxies suggested that there are star-to-star scatters of [Zn/Fe] ratios larger than those of [Mg/Fe] ratios. Skúladóttir et al. [24] showed that there might exist the scatters of [Zn/Fe] at $\gtrsim$ −1.8. They implied that the scatters might be larger than the observational errors. They suggested that the very low ratio of [Zn/Fe] might be an indicator for the contribution of pair-instability supernovae. However, they also pointed out that the data suffered the low signal-to-noise ratios.

Figure 4.3 compares the standard deviations of [Zn/Fe] and [Mg/Fe] as a function of [Fe/H]. As shown in this figure, there is no difference between scatters of [Mg/Fe] and [Zn/Fe] ratios at [Fe/H] > −2.5. This result suggests that the scatters seen in the observations due to the observational errors. If the scatters in the observations are real, these results imply that there is another source of Zn contributed at higher metallicity. At [Fe/H] < −2.5, electron-capture supernovae make scatters of [Zn/Fe] ratios larger than those of [Mg/Fe] ratios.

The ratios of [Zn/Fe] are constant ($\sim$ −0.4) at [Fe/H] $\gtrsim$ −1 (Fig. 4.1). At higher metallicity, there are increasing contributions of core-collapse supernovae (Fig. 2.1). This is due to the yield of Zn by the neutron-capture process in C and He burning.

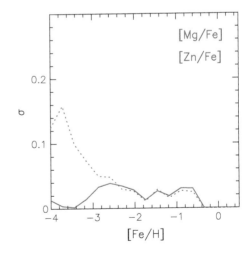

Fig. 4.3 Standard deviations (σ) of [Zn/Fe] ratios (the blue dotted curve) and [Mg/Fe] ratios (the red solid curve) and as functions of [Fe/H] in model G (Fig. 10 of [9], reproduced under the terms of Creative Commons Attribution 3.0 license (https://creativecommons.org/licenses/by/3.0/))

4.3   Mass Ranges of Electron-Capture Supernovae

The mass ranges of electron-capture supernovae affect the enrichment of Zn in galaxies. Figure 4.4 represents [Zn/Fe] ratios as functions of [Fe/H] computed in models with different mass ranges of electron-capture supernovae. In model H (Fig. 4.4a), the mass range predicted by the stellar evolution calculation of [18] is assumed. In model I (Fig. 4.4b), we assume stars with 8.5 to 9.0 $M_\odot$ cause electron-capture supernovae. According to this figure, the ratios of [Zn/Fe] in models H and I are higher than those of model G. This is because models H and I adopt wider mass range than model G. The fractions of electron-capture supernovae compared to all core-collapse supernovae are 3.1% (model G), 38.2% (model H), and 6.9% (model I) at [Fe/H] = -3. In the model I, the mean [Zn/Fe] ratios are consistent with the observations at [Fe/H] $\lesssim -2.5$. This result implies that the decreasing trend of [Zn/Fe] toward

Fig. 4.4 Same as Fig. 4.1, but for models with different mass ranges of electron-capture supernovae. Model H **a** assumes mass ranges predicted by the stellar evolution calculation of Poelarends [18]. Model I **b** adopts the constant mass range from 8.5 to 9.0 $M_\odot$

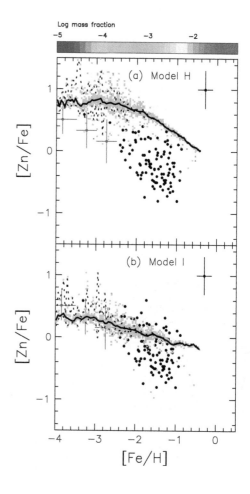

higher metallicity in [Zn/Fe] ratios can be reproduced if the rates of electron-capture supernovae are lower at higher metallicity.

There are several possibilities which can affect the rates of the nucleosynthetic events of Zn. In close binaries, stripped-envelope stars can cause electron-capture supernovae [13, 27, 28]. Several studies predicted that the mass ranges of progenitors of electron-capture supernovae would be wider in binaries [16, 19]. A large amount of iron-peak elements was predicted to be synthesized in the ultra-stripped supernovae [37]. Wanajo et al. [35] showed that iron core-collapse supernovae in the low mass end also synthesized similar nucleosynthetic patterns. These effects could increase the rates of production of Zn.

4.4 Effects of Electron-Capture Supernovae and Hypernovae on the Enrichment of Zn

In this chapter, we consider electron-capture supernovae, hypernovae, and normal core-collapse supernovae as the astrophysical sites of Zn. According to Fig. 2.1, electron-capture supernovae and hypernovae have higher ratios of [Zn/Fe] than those of normal core-collapse supernovae. This result suggests that electron-capture supernovae and hypernovae can be the main astrophysical sites of Zn. However, the contributions of these supernovae on the enrichment of Zn are not clarified. In this section, the contribution of each site of Zn is separately discussed.

Figure 4.5 represents [Zn/Fe] ratios as functions of [Fe/H] computed by models without the effects of electron-capture supernovae. Model J (Fig. 4.5a) assumes 50% of stars more massive than 20 $M_\odot$ explode as hypernovae following Kobayashi et al. [11]. As shown in Fig. 4.5a, there are flat [Zn/Fe] ratios and no stars with [Zn/Fe] $\gtrsim 0.5$ at low metallicity. Model K assumes ten times lower rates of hypernovae than those of model J. Rates of hypernovae in model K are taken from the rates estimated from long gamma-ray bursts [7, 17]. Figure 4.5b shows significantly lower computed [Zn/Fe] ratios than the observed ones. These results suggest that hypernovae cannot fully explain the observed [Zn/Fe] ratios.

Figure 4.6 highlights the contribution of electron-capture supernovae on the enrichment of Zn. There is no contribution of hypernovae in models L and M. As shown in this figure, the ratios of [Zn/Fe] are lower than the observation while model M produces enough amount of Zn to explain the observations at [Fe/H] < -2.5. Higher [Zn/Fe] ratios in model M than those in model L are because model M adopts wider mass ranges of electron-capture supernovae (Table 4.1). These results suggest that the observations of [Zn/Fe] ratios can be reproduced without the effects of hypernovae if the rates of electron-capture supernovae are high enough at low metallicity.

The relation between Zn and trans-iron elements such as Sr, Y, and Zr may be another indicator to constrain the astrophysical sites of Zn. Wanajo et al. [35] reported that electron-capture supernovae synthesized a significant amount of elements from

Fig. 4.5 Same as Fig. 4.1, but for models with no contribution of electron-capture supernovae. Models J **a** and K **b** respectively assume 50% and 5% of stars with more massive than 20 $M_\odot$ explode as hypernovae

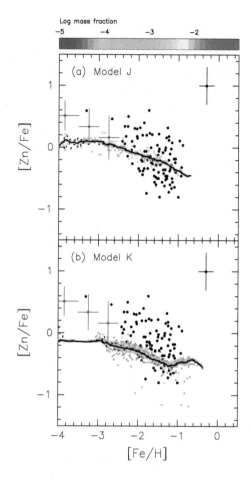

Zn to Zr. Aoki et al. [1] showed that there were variations in the abundances of light neutron-capture elements in metal-poor stars. These variations could be explained by the diversity in the electron-fraction and the mass of proto-neutron stars [33, 34]. Hirai et al. [8] found that the ratios of [Sr/Zn] could be explained by the electron-capture supernovae and neutron star mergers.

4.5 Effects of Yields of Zn and Dependence on the Resolution

Yields of supernovae can vary among different models (e.g., [25]). Figure 4.7 shows [Zn/Fe] as functions of [Fe/H] in models adopting different yields of core-collapse supernovae. Model G adopts yields of Nomoto et al. [14] while model N adopts yields

Fig. 4.6 Same as Fig. 4.1, but for models L **a** and M **b**

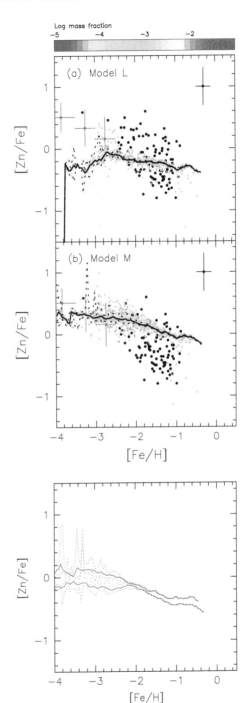

Fig. 4.7 [Zn/Fe] as functions of [Fe/H] in models assuming different yields of core-collapse supernovae (Fig. 14 of [9], reproduced under the terms of Creative Commons Attribution 3.0 license (https://creativecommons. org/licenses/by/3.0/)). Red and blue curves represent models G and N, respectively. Solid and dotted curves mean the same as Fig. 4.1

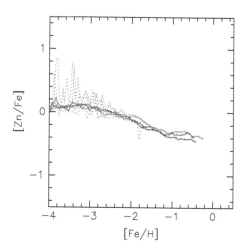

Fig. 4.8 Same as Fig. 4.7, but for models G (red curves), O (green curves), and P (blue curves) (Fig. 15 of [9], reproduced under the terms of Creative Commons Attribution 3.0 license (https://creativecommons. org/licenses/by/3.0/))

of Chieffi and Limongi [2]. As shown in this figure, the [Zn/Fe] ratios in model N are slightly lower than those in model G. However, this difference is much smaller than the uncertainties of the rates of electron-capture supernovae. This result means that the effects of yields are not significant compared to the rates.

Resolution of the simulation may also affect the results. Figure 4.8 shows [Zn/Fe] as functions of [Fe/H] in models with different resolution (see Table 4.1 for the number of particles of each model). As shown in this figure, scatters and median value of [Zn/Fe] ratios are not significantly affected by the resolution. This result means that the resolution of the simulation does not significantly affect the results.

4.6 Summary

The enrichment of Zn in dwarf galaxies was studied in this chapter. The effects of electron-capture supernovae were newly implemented to the simulation. We found that electron-capture supernovae can form stars which have [Zn/Fe] $\gtrsim$ 0.5 (Fig. 4.1). These stars are formed within two billion years from the beginning of the simulation. In this phase, the spatial distribution of metals is still inhomogeneous.

At higher metallicity, the scatters of [Zn/Fe] ratios are not significantly different from [Mg/Fe] ratios. This result suggests that the scatters of [Zn/Fe] ratios at [Fe/H] $\gtrsim$ −2.5 are caused by the contribution of type Ia supernovae. Stars with [Zn/Fe] < −0.5 in our models are formed at less than four billion years from the start of the simulation (Fig. 4.2).

The contribution of different sources of Zn has been discussed in this chapter. It was not possible to reproduce stars with high [Zn/Fe] ratios without considering the effects of electron-capture supernovae (Fig. 4.5). We found that mass ranges of electron-capture supernovae significantly affect the ratios of [Zn/Fe] (Figs. 4.4 and 4.6). From these results, we can conclude that electron-capture supernovae or iron core-collapse supernovae from low mass progenitors are important contributors to the enrichment of Zn in galaxies.

References

1. Aoki M, Ishimaru Y, Aoki W, Wanajo S (2017) Diversity of abundance patterns of light neutron-capture elements in very-metal-poor stars. Astrophys J 837:8
2. Chieffi A, Limongi M (2004) Explosive yields of massive stars from $Z = 0$ to $Z = Z_{solar}$. Astrophys J 608:405–410
3. Doherty CL, Gil-Pons P, Siess L, Lattanzio JC, Lau HHB (2015) Super- and massive AGB stars - IV. Final fates—initial-to-final mass relation. Mon Not R Astron Soc 446:2599–2612
4. François P, Matteucci F, Cayrel R, Spite M, Spite F, Chiappini C (2004) The evolution of the Milky Way from its earliest phases: constraints on stellar nucleosynthesis. Astron Astrophys 421:613–621
5. Geisler D, Smith VV, Wallerstein G, Gonzalez G, Charbonnel C (2005) "Sculptor-ing" the galaxy? The chemical compositions of red giants in the sculptor dwarf spheroidal galaxy. Astron J 129:1428–1442
6. Goswami A, Prantzos N (2000) Abundance evolution of intermediate mass elements (C to Zn) in the MilkyWay halo and disk. Astron Astrophys 359:191–212
7. Guetta D, Della Valle M (2007) On the rates of gamma-ray bursts and type Ib/c supernovae. Astrophys J Lett 657:L73–L76
8. Hirai Y, Wanajo S, Saitoh TR (2019) Contribution of electron-capture supernovae and neutron star mergers to the enrichment of strontium in dwarf galaxies. Astrophys J submitted
9. Hirai Y, Saitoh TR, Ishimaru Y, Wanajo S (2018) Enrichment of zinc in galactic chemodynamical evolution models. Astrophys J 855(63):63
10. Ishimaru Y, Wanajo S, Prantzos N (2006) Chemical evolution of C-Zn and r-process elements produced by the first generation stars. In: International Symposium on Nuclear Astrophysics—Nuclei in the Cosmos, p 37.1
11. Kobayashi C, Umeda H, Nomoto K, Tominaga N, Ohkubo T (2006) Galactic chemical evolution: carbon through zinc. Astrophys J 653:1145–1171
12. Matteucci F, Raiteri CM, Busson M, Gallino R, Gratton R (1993) Constraints on the nucleosynthesis of Cu and Zn from models of chemical evolutino of the galaxy. Astron Astrophys 272:421
13. Moriya TJ, Eldridge JJ (2016) Rapidly evolving faint transients from stripped-envelope electron-capture supernovae. Mon Not R Astron Soc 461:2155–2161
14. Nomoto K, Kobayashi C, Tominaga N (2013) Nucleosynthesis in stars and the chemical enrichment of galaxies. Annu Rev Astron Astrophys 51:457–509
15. Nomoto K, Hashimoto M, Tsujimoto T, Thielemann F-K, Kishimoto N, Kubo Y, Nakasato N (1997) Nucleosynthesis in type II supernovae. Nucl Phys A 616:79–90
16. Podsiadlowski P, Langer N, Poelarends AJT, Rappaport S, Heger A, Pfahl E (2004a) The effects of binary evolution on the dynamics of core collapse and neutron star kicks. Astrophys J 612:1044–1051
17. Podsiadlowski P, Mazzali PA, Nomoto K, Lazzati D, Cappellaro E (2004b) The rates of hypernovae and gamma-ray bursts: implications for their progenitors. Astrophys J Lett 607:L17–L20
18. Poelarends AJT (2007) Stellar evolution on the borderline of white dwarf and neutron star formation. PhD thesis, Utrecht University
19. Poelarends AJT, Wurtz S, Tarka J, Cole Adams L, Hills ST (2017) Electron capture supernovae from close binary systems. Astrophys J 850:197

20. Saito Y-J, Takada-Hidai M, Honda S, Takeda Y (2009) Chemical evolution of zinc in the galaxy. Publ Astron Soc Jpn 61:549–561
21. Shetrone M, Venn KA, Tolstoy E, Primas F, Hill V, Kaufer A (2003) VLT/UVES abundances in four nearby dwarf spheroidal galaxies I. Nucleosynthesis and abundance ratios. Astron J 125:684–706
22. Simon JD, Jacobson HR, Frebel A, Thompson IB, Adams JJ, Shectman SA (2015) Chemical signatures of the first supernovae in the sculptor dwarf spheroidal galaxy. Astrophys J 802:93
23. Skúladóttir Á, Tolstoy E, Salvadori S, Hill V, Pettini M, Shetrone MD, Starkenburg E (2015) The first carbon-enhanced metal-poor star found in the sculptor dwarf spheroidal. Astron Astrophys 574:A129
24. Skúladóttir Á, Tolstoy E, Salvadori S, Hill V, Pettini M (2017) Zinc abundances in the sculptor dwarf spheroidal galaxy. Astron Astrophys 606:A71
25. Sukhbold T, Ertl T, Woosley SE, Brown JM, Janka H-T (2016) Corecollapse supernovae from 9 to 120 solar masses based on neutrino-powered explosions. Astrophys J 821(38):38
26. Tanaka M, Tominaga N, Nomoto K, Valenti S, Sahu DK, Minezaki T, Yoshii Y, Yoshida M, Anupama GC, Benetti S, Chincarini G, Della Valle M, Mazzali PA, Pian E (2009) Type Ib supernova 2008D associated with the luminous X-ray transient 080109: an energetic explosion of a massive helium star. Astrophys J 692:1131–1142
27. Tauris TM, Langer N, Podsiadlowski P (2015) Ultra-stripped supernovae: progenitors and fate. Mon Not R Astron Soc 451:2123–2144
28. Tauris TM, Langer N, Moriya TJ, Podsiadlowski P, Yoon S-C, Blinnikov SI (2013) Ultra-stripped type Ic supernovae from close binary evolution. Astrophys J Lett 778:L23
29. Timmes FX, Woosley SE, Weaver TA (1995) Galactic chemical evolution: hydrogen through zinc. Astrophys J 98:617–658
30. Tominaga N, Umeda H, Nomoto K (2007) Supernova nucleosynthesis in population III 13–50 msolar stars and abundance patterns of extremely metalpoor stars. Astrophys J 660:516–540
31. Umeda H, Nomoto K (2002) Nucleosynthesis of zinc and iron peak elements in population III type II supernovae: comparison with abundances of very metal poor halo stars. Astrophys J 565:385–404
32. Umeda H, Nomoto K (2005) Variations in the abundance pattern of extremely metal-poor stars and nucleosynthesis in population III supernovae. Astrophys J 619:427–445
33. Wanajo S (2013) The r-process in proto-neutron-star wind revisited. Astrophys J Lett 770:L22
34. Wanajo S, Janka H-T, Müller B (2011) Electron-capture supernovae as the origin of elements beyond iron. Astrophys J Lett 726:L15
35. Wanajo S, Müller B, Janka H-T, Heger A (2018) Nucleosynthesis in the innermost ejecta of neutrino-driven supernova explosions in two dimensions. Astrophys J 852:40
36. Woosley SE, Weaver TA (1995) The evolution and explosion of massive stars II. Explosive hydrodynamics and nucleosynthesis. Astrophys J 101:181
37. Yoshida T, Suwa Y, Umeda H, Shibata M, Takahashi K (2017) Explosive nucleosynthesis of ultra-stripped type Ic supernovae: application to light trans-iron elements. Mon Not R Astron Soc 471:4275–4285

Chapter 5
Enrichment of r-Process Elements in Isolated Dwarf Galaxies

Abstract The abundances of r-process elements in stars can be a signature for the astrophysical sites of the r-process elements and star formation histories of galaxies. Observations have confirmed that there are star-to-star scatters of the ratios r-process elements to iron at low metallicity in the Milky Way halo. However, the enrichment history of r-process elements has not yet been clarified. In this chapter, we have computed the enrichment of r-process elements using a series of N-body/hydrodynamic simulations of dwarf galaxies with various initial conditions. We find that neutron star mergers can enrich r-process elements at [Fe/H] $\lesssim -2.5$ if the star formation efficiency of the galaxy is less than 0.01 per billion years. In such galaxies, the mean metallicity is almost constant over several hundred million years. On the other hand, the r-process elements are produced at higher metallicity in galaxies with higher star formation efficiencies. The star-to-star scatters of the ratios of r-process elements to iron mainly come from the inhomogeneous spatial metallicity distribution in galaxies. The results demonstrate that it will be possible to estimate the early star formation histories of galaxies by using the abundances of r-process elements (Contents in this chapter have been in part published in Hirai et al. (Astrophys J 814:41, [19]), Hirai et al. (Mon Not Royal Astron Soc 466:2474-2487, [20]) reproduced permission of the AAS and Oxford University Press).

5.1 Enrichment of r-Process Elements in the Local Group

Stellar abundances of r-process elements help us understand the origin of these elements and the early star formation histories of galaxies. Observations with high-dispersion spectroscopy have shown that there are large star-to-star variations of r-process elements to iron ([r/Fe]) at low metallicity in the Milky Way halo (Fig. 1.10). These observations indicate that the astrophysical sites of r-process elements are rarer events than all supernovae.

The Local Group dwarf galaxies also have r-process elements at low metallicity. Scatters of the ratios of r-process elements to iron in dwarf galaxies are within the scatters in the Milky Way halo stars. Although it is difficult to determine the degree of

© Springer Nature Singapore Pte Ltd. 2019
Y. Hirai, *Understanding the Enrichment of Heavy Elements by the Chemodynamical Evolution Models of Dwarf Galaxies*, Springer Theses, https://doi.org/10.1007/978-981-13-7884-3_5

scatters of [*r*/Fe] ratios, the scatters in each dwarf galaxy tend to be smaller than those in the Milky Way halo stars. The *r*-process enhanced stars at [Fe/H] $\lesssim -2.5$ have not yet found in dwarf spheroidal galaxies (e.g., [15]). On the other hand, several stars in the Reticulum II ultra-faint dwarf galaxy [22, 23, 35] show enhanced *r*-process abundances.

Astrophysical sites of *r*-process are not well understood. Binary neutron star mergers are the most promising astrophysical site of the *r*-process (see Chap. 1). The ejecta of neutron star mergers are sufficiently neutron-rich to synthesize elements with the mass number larger than 110 (e.g., [4, 16, 18, 27, 33, 36, 43, 53]). On 2017, multi-messenger observations detect the afterglows of the neutron star merger, GW170817 (see Chap. 1 and references therein). The decay of radioactive elements synthesized in the neutron star merger can explain this afterglow (e.g., [47]). This observation can be direct evidence of the *r*-process in neutron star mergers.

Galactic chemical evolution studies suggested that it was difficult to explain observations by neutron star mergers (e.g., [3, 30]). Significant fractions of binary neutron stars were estimated to have merger timescale longer than 100 million years (e.g., [5, 6, 14, 24, 29]). In addition, the rate of neutron star mergers was estimated to be ~0.1–1 % of all core-collapse supernovae (e.g., [1]). Argast et al. [3] pointed out that the metallicity in their model was too high to explain the observations at the time when the first neutron star merger occurred due to the low rates and long merger times of neutron star mergers.

Recently, chemical evolution studies proposed several solutions to this problem [8, 11, 19, 21, 25, 26, 31, 32, 40, 48–51, 54]. [21] showed that neutron star mergers produced *r*-process elements in lower metallicity if they assumed star formation efficiencies were lower in less massive sub-halos. Star formation efficiencies of galaxies are highly related to the dynamical evolution and cooling and heating of the interstellar medium. The chemodynamical approach is thus necessary to self-consistently simulate the evolution of small halos.

In this chapter, we performed a series of chemodynamical simulations of dwarf galaxies assuming neutron star mergers are the major astrophysical site of the *r*-process. In Sect. 5.2, we show that neutron star mergers can produce *r*-process elements at low metallicity in a dwarf galaxy. Sections 5.3 and 5.4 show the dependence on the initial conditions of halos. Section 5.7 discusses the relation to the Milky Way halo formation. In Sect. 5.8, we summarize the results in this chapter. Table 5.1 lists models adopted in this chapter.

5.2   Enrichment of *r*-Process Elements in a Dwarf Galaxy

In this section, we show the enrichment of *r*-process elements in an isolated dwarf galaxy model. Figure 5.1 shows [Eu/Fe] as functions of [Fe/H] in model m000. This model includes the effects of metal mixing in star-forming regions. In model m000, there are star-to-star scatters of [Eu/Fe] at [Fe/H] $\lesssim -2.5$ even if we assume neutron star mergers with merger times of 100 million years.

Table 5.1 List of our models discussed in this chapter

Model	M_{tot} ($10^8 M_\odot$)	ρ_c ($10^7 M_\odot$ kpc^{-3})	m_{DM} ($10^3 M_\odot$)	m_{gas} ($10^3 M_\odot$)	r_{max} (kpc)	r_c (kpc)	n_{th} cm^{-3}	ε_{SN} (erg)	t_{NSM} (million years)	f_{NSM}
m000	7.0	0.5	2.3	0.4	7.1	1.0	100	1.0×10^{51}	100	0.01
mExt	7.0	0.5	2.3	0.4	7.1	1.0	0.1	3.0×10^{49}	100	0.01
A	3.5	0.5	1.1	0.2	7.1	1.0	100	1.0×10^{51}	100	0.01
B	3.5	1.5	1.1	0.2	4.9	0.7	100	1.0×10^{51}	100	0.01
C	3.5	10.0	1.1	0.2	2.6	0.4	100	1.0×10^{51}	100	0.01
D	3.5	10.0	1.1	0.2	1.0	1.0	100	1.0×10^{51}	100	0.01
E	7.0	0.5	2.3	0.4	8.9	1.3	100	1.0×10^{51}	100	0.01
F	35.0	0.5	11.3	2.0	15.3	2.2	100	1.0×10^{51}	100	0.01
mt10	7.0	0.5	2.3	0.4	7.1	1.0	100	1.0×10^{51}	10	0.01
mt500	7.0	0.5	2.3	0.4	7.1	1.0	100	1.0×10^{51}	500	0.01
A10	3.5	0.5	1.1	0.2	7.1	1.0	100	1.0×10^{51}	10	0.01
B10	3.5	1.5	1.1	0.2	4.9	0.7	100	1.0×10^{51}	10	0.01
C10	3.5	10.0	1.1	0.2	2.6	0.4	100	1.0×10^{51}	10	0.01
D10	3.5	10.0	1.1	0.2	1.0	1.0	100	1.0×10^{51}	10	0.01
A500	3.5	0.5	1.1	0.2	7.1	1.0	100	1.0×10^{51}	500	0.01
B500	3.5	1.5	1.1	0.2	4.9	0.7	100	1.0×10^{51}	500	0.01
C500	3.5	10.0	1.1	0.2	2.6	0.4	100	1.0×10^{51}	500	0.01
D500	3.5	10.0	1.1	0.2	1.0	1.0	100	1.0×10^{51}	500	0.01
mt0.001	7.0	0.5	2.3	0.4	7.1	1.0	100	1.0×10^{51}	100	0.001
mr0.1	7.0	0.5	2.3	0.4	7.1	1.0	100	1.0×10^{51}	100	0.1

From left to right, we show the name of models, the total mass of halos (M_{tot}), the initial central densities of halos (ρ_c), the mass of one dark matter particle (m_{DM}), the mass of one gas particle (m_{gas}), the maximum radius of halos (r_{max}), the core radius of halos (r_c), the threshold density for star formation (n_{th}), the thermal energy of supernova feedback (ε_{SN}), merger times of neutron star mergers (t_{NSM}), and fraction of neutron star mergers in stars with 8–20 $M_\odot$ (f_{NSM})

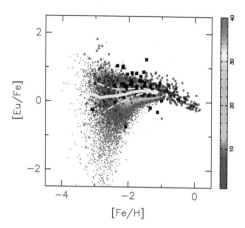

Fig. 5.1 [Eu/Fe] as functions of [Fe/H] in model m000 (Fig. 7 of [19], reproduced by permission of the AAS). The number of computed stars is described in color from 0 (purple) to 20 (red). The yellow solid, and dash-dotted curves represent the median value, the first and third quartiles of the results, respectively. Squares and circles denote the observations of the Ursa Minor, Sculptor, Leo I, Draco, Carina dwarf spheroidal galaxies and the Milky Way halo (SAGA database, [44–46, 55]). Computed stars are plotted in 0.5 kpc from the center of the galaxy to be comparable to the observations

This result is different from that of Argast et al. [3], which shows difficulty in explaining the observations of *r*-process elements in stars with [Fe/H] $\lesssim -2.5$. The significant difference in model m000 and models of Argast et al. [3] is the star formation efficiencies in the early phases of galaxy evolution. Here we define the star formation efficiencies as star formation rates divided by gas mass. In models of Argast et al. [3], the typical star formation efficiency corresponds to ~ 1 Gyr^{-1}. On the other hand, model m000 has star formation efficiency of ~ 0.01 Gyr^{-1} during the first one billion years. This efficiency is consistent with models of Ishimaru et al. [21]. This is because supernova feedback efficiently suppresses the star formation in dwarf galaxies.

Figure 5.2 shows the reason why it is possible to produce stars with *r*-process elements in [Fe/H] $\lesssim -2.5$. According to this figure, metallicity is constant at $\lesssim$ 300 million years from the beginning of the significant star formation. This is due to the low star formation efficiency and inhomogeneous spatial metallicity distribution in the early phase. The yields of each supernova determine the metallicity of stars formed in this period. At $t > 300$ million years, spatial metallicity distribution becomes homogeneous. In this phase, metallicity increases proportional to the number of supernovae.

The difference of star formation histories seen between models m000 and mExt (Fig. 3.9) has enormous impacts on the enrichment of *r*-process elements. Figure 5.3 shows [Eu/Fe] as functions of [Fe/H] computed on model mExt. The energy of this model is 2 dex lower than that of model m000. In addition to the energy of supernova feedback, the threshold density for star formation is lower than that of

Fig. 5.2 Time evolution of [Fe/H] in model m000 (Fig. 12 of [19], reproduced by permission of the AAS). The number of computed stars is described in color from 0 (purple) to 20 (red). The horizontal axis depicts time from the start of the major star formation (≈600 million years from the beginning of the simulation). Black curve represents the mean value of [Fe/H]

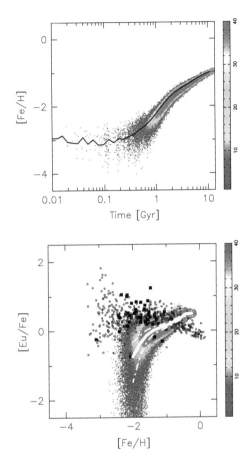

Fig. 5.3 Same as Fig. 5.1 but for model mExt (Fig. 10 of [19], reproduced by permission of the AAS)

model m000. This difference induces higher star formation rates in the early phase. The star formation efficiency in this model is consistent with models of Argast et al. [3]. In the following section, we systematically discuss the effects of star formation histories on the enrichment of *r*-process elements.

5.3   Effects of the Initial Central Density of Halos

The central densities of halos significantly affect the star formation histories of galaxies (Fig. 3.6). Chemical evolution of galaxies depends on the star formation histories. Figure 5.4 depicts the evolution of the average stellar metallicity in models A, B, C, and D. Hereafter, we define the mean metallicity at 100 million years from the beginning of the simulation as $[Fe/H]_0$ (crosses in Fig. 5.4). The value of $[Fe/H]_0$ indicates metallicity of galaxies when the first neutron star merger occurs. The values

Fig. 5.4 Time evolution of mean stellar metallicities in models A (red), B (green), C (blue), and D (orange, Fig. 9 of [20], reproduced by permission of the Oxford University Press). Vertical solid lines indicate dispersion of metallicity in 2σ. The values of $[Fe/H]_0$ (see text for definition) are indicated by the crosses. Vertical dotted lines mean the time one billion years from the beginning of the star formation

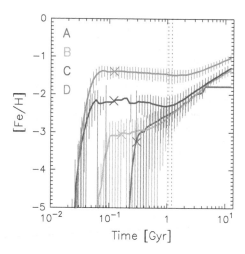

of $[Fe/H]_0$ are $[Fe/H]_0 = -3.2$ (model A), -3.0 (model B), -2.2 (model C), and -1.4 (model D). This result means that models with a higher central density of halos have higher values of $[Fe/H]_0$. This is because star formation rates are higher in models with higher central densities (Fig. 3.6). The mean $[Fe/H]$ of models C and D does not increase from 0.1 to 1 billion years from the beginning of the simulation because star formation rates are suppressed during this time.

The ratios of $[Eu/Fe]$ reflect the early evolution of $[Fe/H]$. Figure 5.5 denotes $[Eu/Fe]$ as functions of $[Fe/H]$ in models A, B, C, and D. As shown in this figure, Eu appears at higher metallicity in models with higher central densities. Models A and B have stars with Eu at $[Fe/H] \sim -3$ while models C and D do not have such stars. Stars with Eu at $\lesssim -3$ may come from galaxies like models A and B. Galaxies similar to models C and D can contribute to the enrichment of *r*-process elements only at $[Fe/H] \gtrsim -2$.

Scatters of $[Eu/Fe]$ ratios become smaller in models with higher central densities. Figure 5.6 compares standard deviation (σ) of $[Eu/Fe]$ ratios as functions of $[Fe/H]$ in models A, B, C, and D. Metallicity distribution tends to be more inhomogeneous in models with lower star formation efficiencies (see Fig. 3.15). The $[Eu/Fe]$ ratios reflect these inhomogeneities.

It is possible to quantify the degree of inhomogeneity of metals by using a metal pollution factor (f_{poll}, [2]),

$$f_{poll} = \frac{M_{poll}}{M_{gas}}, \tag{5.1}$$

where M_{poll} and M_{gas} are the gas mass polluted by metals and the total mass of gases in a galaxy, respectively. In this work, we assume metals are mixed in a star-forming region. Therefore, we can define M_{poll} as

$$M_{poll} = N_{SN}M_{sw} + N_{\star}M_{mix}, \tag{5.2}$$

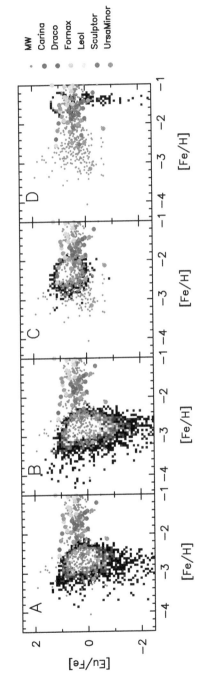

Fig. 5.5 [Eu/Fe] ratios as functions of [Fe/H] (the right most panel, Fig. 10 of [20], reproduced by permission of the Oxford University Press). From left to right, panels represent models A, B, C, and D. The results are plotted at one billion years from the beginning of the star formation. The number of stars in each bin is described as gray scales. The colored dots represent observed stellar abundances of the Milky Way halo (sky-blue [34]), Ursa Minor (orange [10, 38, 42]), Sculptor (magenta [17, 41]), Leo I (cyan [41]), Fornax (green [28]), Draco (blue [9, 42]), and Carina (red [28, 41, 52]). All data are compiled using SAGA database [44–46, 55]

Fig. 5.6 Standard deviation (σ) of [Eu/Fe] ratios as functions of [Fe/H] at one billion years from the start of the star formation in models A (red), B (green), C (blue), and D (orange, Fig. 11 of [20], reproduced by permission of the Oxford University Press)

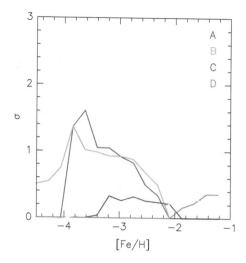

where N_{SN}, M_{sw}, $N_\star$, and M_{mix} are the number of supernovae, the gas mass inside the SPH kernel, the number of star particles, the gas mass in a star-forming region, respectively. If the value of f_{poll} is larger than unity, the whole space is polluted by metals. In other words, there are no small scale spatial variations of metallicity. On the other hand, there are inhomogeneities in the metallicity distribution if the value of f_{poll} is less than unity. The calculated values of f_{poll} in models A, B, C, and D are 0.71, 0.26, 2.5, and 41.1, respectively, at 200 million years from the start of the star formation. This result means that the metallicity distribution is still inhomogeneous in models A and B while metals in models C and D are already homogenized at this phase.

The Local Group dwarf spheroidal galaxies such as Sculptor, Fornax, Draco, and Carina have r-process elements at [Fe/H] $\lesssim -2.5$ (e.g., [15]). The star formation rates in Sculptor and Fornax computed from the color-magnitude diagrams are $\sim 10^{-3}\, M_\odot \mathrm{yr}^{-1}$ [12, 13]. These values are similar to models A and B. Star formation efficiencies computed in models A and B are ~ 0.01 per one billion years. This result suggests that star formation efficiencies of galaxies should be less than 0.01 per one billion years to have Eu at [Fe/H] $\lesssim -2.5$.

If a galaxy has r-process elements only at high metallicity, such galaxies may have efficiently formed stars during the early phases of galaxy evolution like models C and D. The dwarf spheroidal galaxy Sagittarius can be a candidate of such galaxies. Although the number of observed stars is not enough, stars with r-process elements at [Fe/H] < -1 have not yet confirmed so far in this galaxy (e.g., [7, 39]). This galaxy may have formed with the higher star formation efficiencies than those of Sculptor or Fornax dwarf spheroidal galaxies.

In this section, we find that Eu appears at higher metallicity in galaxies with higher initial central densities of halos. The initial central densities control star formation efficiencies of galaxies. The results demonstrate that the r-process elements can be an indicator of early star formation efficiencies of galaxies. It is important to determine the metallicity at which Eu appears in each galaxy to understand the early star formation histories.

5.4 Effects of the Total Mass of Halos

The total mass of halos also affects the final properties of galaxies (Chap. 3). Star formation histories vary among models with different total mass (Fig. 3.7). In this section, we show the effects of the total mass of halos.

Figure 5.7 shows [Fe/H] ratios as functions of time. All models in this figure have $[Fe/H]_0 < -3.1$ because of the low star formation rates at the early phase of the evolution. As shown in this figure, there are large scatters of [Fe/H] at the time less than one billion years. Models A, E, and F have $f_{poll} = 0.33, 0.0023$, and 0.011, respectively, at 100 million years from the beginning of the star formation. This result means that the total mass does not significantly affect the enrichment of Fe in the early phases of galaxy evolution.

The total mass of halos has little effects on the [Eu/Fe] ratios at low metallicity. Figure 5.8 represents [Eu/Fe] as functions of [Fe/H] computed in models A, E, and F. As shown in this figure, all models have stars with Eu at low metallicity. This is due to slow chemical evolution in all of these models. It takes over one billion years to exceed $[Fe/H] = -3$ in these models (Fig. 5.4), i.e., there is plenty of time for binary neutron stars to merge at low metallicity.

Morel F has few stars at one billion years from the beginning of the simulation because star formation rates are suppressed below 10^{-5} $M_\odot$ yr^{-1} in this model (Fig. 3.7). Most of the stars in model F are formed after one billion years from the start of the simulation. We thus also plot the results at the time of three billion years.

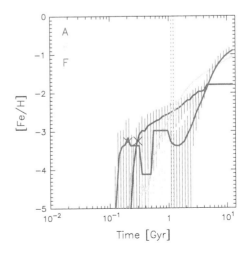

Fig. 5.7 Same as Fig. 5.4, but for models A (red), E (cyan), and F (magenta, Fig. 13 of [20], reproduced by permission of the Oxford University Press)

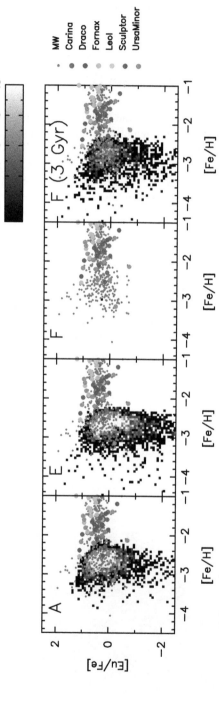

Fig. 5.8 Same as Fig. 5.5, but for models A (the leftmost panel), E (the second left panel), F (the second right panel) at one billion years from the beginning of the simulation and model F (the rightmost panel) at the time of three billion years (Fig. 14 of [20], reproduced by permission of the Oxford University Press)

Fig. 5.9 Same as Fig. 5.6, but for models A (red), E (cyan), and F (magenta, Fig. 15 of [20], reproduced by permission of the Oxford University Press)

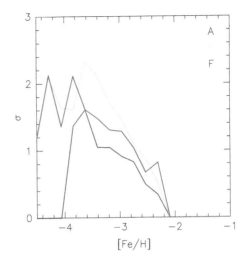

Scatters of [Eu/Fe] are similar among models A, E, and F. Figure 5.9 compares standard deviations of the ratios of [Eu/Fe] in these models. At [Fe/H] $\lesssim -3$, all models have $\sigma > 1$. This result is expected from the large scatters seen in Fig. 5.7. From these results, the total mass of halos does not significantly affect the early enrichment of r-process elements in dwarf galaxies.

5.5   Merger Times of Neutron Star Mergers

Merger times of neutron star mergers may affect the enrichment of r-process elements. In this section, models assuming different neutron star merger times are compared. Figure 5.10 shows [Eu/Fe] as functions of [Fe/H] computed in models with different merger times of neutron star mergers. Figure 5.10a represents the result of model mt10. This model adopts a neutron star merger time of 10 million years. Shorter timescales of neutron star mergers preferred in previous studies (e.g., [31, 49]). As shown in this figure, there are stars with Eu at [Fe/H] $\lesssim -2.5$. Figure 5.10b denotes the results computed on model mt500, which assumes neutron star merger times of 500 million years. Although the effect of delayed production of Eu can be seen at [Fe/H] ~ -2, there are stars with Eu at [Fe/H] $\lesssim -2.5$. This result suggests that neutron star mergers with merger times shorter than 500 million years are possible to contribute to the enrichment of r-process elements in low metallicity.

The effect of merger times of neutron star mergers is marginal in models with different central densities. Figure 5.11 shows [Eu/Fe] as functions of [Fe/H] computed in models A_{10}, B_{10}, C_{10}, and D_{10}. These models assume neutron star mergers

Fig. 5.10 Same as Fig. 5.1, but for models **a** mt10 and **b** mt500 (Fig. 11 of [19], reproduced by permission of the AAS)

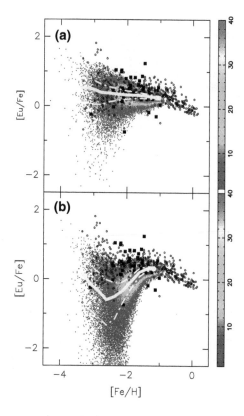

with the merger time of 10 million years. Models A_{10} and B_{10} have similar results with models A and B. Models C_{10} and D_{10} have larger fraction of stars with [Eu/Fe] <0 than models C and D. This is because spatial metallicity distribution is more inhomogeneous at the earlier phase of the galaxy evolution. However, they cannot explain Eu abundances at [Fe/H] $\lesssim -3$. These results suggest that models C_{10} and D_{10} proceed chemical evolution too fast to reproduce the observations of Eu.

Figure 5.12 shows [Eu/Fe] as functions of [Fe/H] computed in models A_{500}, B_{500}, C_{500}, and D_{500}. The scatters of [Fe/H] show over one dex at 500 million years from the start of the star formation in models A and B (Fig. 5.7). This cause scatters of [Eu/Fe] in models A_{500} and B_{500}. Models C_{500} and D_{500} show similar results with models C and D. These results suggest that merger times of neutron star mergers from 10 to 500 million years do not largely affect the results in these models.

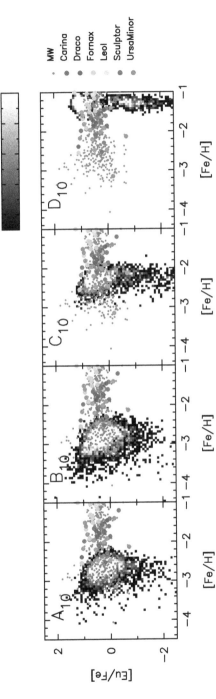

Fig. 5.11 Same as Fig. 5.5, but for models with the neutron star merger time of 10 million years (Fig. 16 of [20], reproduced by permission of the Oxford University Press). From left to right, panels represents models A_{10}, B_{10}, C_{10}, and D_{10}

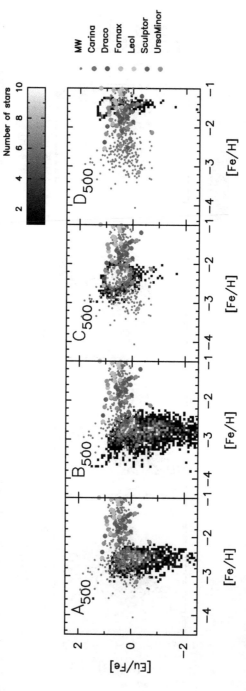

Fig. 5.12 Same as Fig. 5.5, but for models with the neutron star merger time of 500 million years (Fig. 17 of [20], reproduced by permission of the Oxford University Press). From left to right, panels represent models A_{500}, B_{500}, C_{500}, and D_{500}

5.6 Rates of Neutron Star Mergers

In this section, we discuss the effects of the rate of neutron star mergers. The rate of neutron star mergers is highly uncertain. The estimated rate of neutron star mergers is 1540^{+3200}_{-1220} Gpc^{-3}yr^{-1} [1]. Rates of neutron star mergers affect the scatters of [Eu/Fe] ratios. In this study, yields are adjusted to reproduce the average value of [Eu/Fe] at [Fe/H] ~ -1. Table 5.2 lists rates and yields adopted here. All parameters are the same except for the rates of neutron star mergers among models mr0.001, m000, and mr0.1.

Table 5.2 List of rates and total ejecta mass of r-process elements

Model	R_{NSM} Gpc^{-3}yr^{-1}	M_r ($M_\odot$)	f_{NSM}
mr0.001	10^2	10^{-1}	0.001
m000	10^3	10^{-2}	0.01
mr0.1	10^4	10^{-3}	0.1

The columns represent the name of models, rates of neutron star mergers (R_{NSM}), yields of all r-process elements (M_r), and the number fraction of neutron star mergers relative to supernovae with the mass range of $8 - 20\ M_\odot$ (f_{NSM})

Fig. 5.13 Same as Fig. 5.1 but for models **a** mr0.001 and **b** mr0.1 (Fig. 13 of [19], reproduced by permission of the AAS)

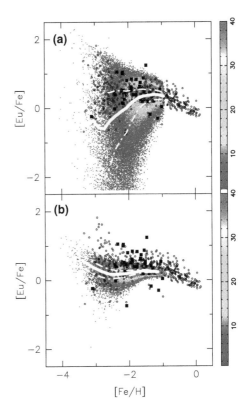

Figure 5.13 shows the effects of the rates of neutron star mergers. According to Fig. 5.13, model mr0.001 shows star-to-star scatters of more than 1 dex even at [Fe/H] ~ -1.5. In contrast, model mr0.1 does not have stars with [Eu/Fe] > 1.5. These results imply that the rates of neutron star mergers significantly affect the scatters of [Eu/Fe] ratios. The estimation of the rates will be improved in the future observations of gravitational wave detectors.

5.7   Implications to the Milky Way Halo Formation

Section 5.2 has shown that models which have low star formation efficiencies ($\sim$0.01 Gyr^{-1}) can explain the observed [Eu/Fe] ratios of extremely metal-poor stars. This result means that star formation efficiencies of sub-halos which contribute to the formation of extremely metal-poor stars with *r*-process elements may have star formation efficiencies of ~ 0.01 Gyr^{-1}.

Assemblies of sub-halos with low star formation efficiencies are not enough to fully explain the observed *r*-process abundances in the Milky Way halo. The mean stellar [Fe/H] of the Milky Way halo is [Fe/H] = -1.6 [37]. At around [Fe/H] = -1.6, the ratios of α-elements to iron are constant, indicating that there is the little contribution of type Ia supernovae at around this metallicity. Type Ia supernovae typically start to contribute to the enrichment of Fe at around one billion years from the beginning of the star formation. As shown in Figs. 5.4 and 5.7, the mean [Fe/H] in models with star formation efficiencies of $\sim$0.01 Gyr^{-1} (models A, B, E, and F) is [Fe/H] $\lesssim -2.5$ at one billion years from the beginning of the star formation. This result suggests that sub-halos with higher star formation efficiencies such as models C and D should contribute to the chemical evolution of the Milky Way halo at [Fe/H] > -2.5.

The speed of chemical evolution would be underestimated in the isolated dwarf galaxy models adopted in this work. In the cosmological context, mergers of galaxies may induce star formation. In Chap. 7, we show the enrichment of *r*-process elements in a cosmological context.

5.8   Summary

In this chapter, we studied the enrichment of *r*-process elements in dwarf galaxies. We found that neutron star mergers with merger times of 100 million years can enrich *r*-process elements at [Fe/H] $\lesssim -2.5$ in galaxies with star formation efficiencies of 0.01 Gyr^{-1} (Fig. 5.1). This is because metallicity is almost constant at $\sim$300 million years from the beginning of the star formation in these models.

The enrichment of r-process elements is affected by the star formation efficiencies of galaxies which are controlled by the dynamical evolution of halos. We found that the initial central density of halos significantly affects the enrichment of r-process elements. The r-process elements appear at higher metallicity in models with higher initial central densities of halos (Fig. 5.5). Scatters of [Eu/Fe] ratios decrease with models with higher initial densities (Fig. 5.6). On the other hand, the total mass of halos does not significantly affect the enrichment of r-process elements (Fig. 5.8). These results suggest that r-process abundances are very sensitive to the early star formation efficiencies of galaxies.

The effects of merger times and rates of neutron star mergers are also examined in this chapter. We found that the [Eu/Fe] ratios are not significantly affected by the merger times from 10 to 500 million years. Neutron star mergers with the merger time of 500 million years can produce r-process elements at [Fe/H] < -2.5 in models with early star formation efficiency of 0.01 Gyr^{-1} (Fig. 5.11). We also found that the rates of neutron star merger may affect the scatters of [Eu/Fe] ratios.

The results in this section suggest that the observed abundances of r-process elements can be explained by neutron star mergers if the Milky Way halo is formed from an assembly of sub-halos with early star formation efficiencies of 0.01 Gyr^{-1}. The result also implies that the abundances of r-process elements can be an indicator of the early star formation histories of galaxies. We expect that future observations of r-process elements in metal-poor stars can constrain the formation and evolution of the Local Group galaxies.

References

1. Abbott BP et al (2017) GW170817: observation of gravitational waves from a binary neutron star inspiral. Phys Rev Lett 119(16):161101
2. Argast D, Samland M, Gerhard OE, Thielemann F-K (2000) Metal-poor halo stars as tracers of ISM mixing processes during halo formation. Astron Astrophys 356:873–887
3. Argast D, Samland M, Thielemann F-K, Qian Y-Z (2004) Neutron star mergers versus core-collapse supernovae as dominant r-process sites in the early Galaxy. Astron Astrophys 416:997–1011
4. Bauswein A, Goriely S, Janka H-T (2013) Systematics of dynamical mass ejection, nucleosynthesis, and radioactively powered electromagnetic signals from neutron-star mergers. Astrophys J 773:78
5. Belczynski K, Kalogera V, Bulik T (2002) A comprehensive study of binary compact objects as gravitational wave sources: evolutionary channels, rates, and physical properties. Astrophys J 572:407–431
6. Belczynski K, Perna R, Bulik T, Kalogera V, Ivanova N, Lamb DQ (2006) A study of compact object mergers as short gamma-ray burst progenitors. Astrophys J 648:1110–1116
7. Bonifacio P, Hill V, Molaro P, Pasquini L, Di Marcantonio P, Santin P (2000) First results of UVES at VLT: abundances in the Sgr dSph. Astron Astrophys 359:663–668
8. Cescutti G, Romano D, Matteucci F, Chiappini C, Hirschi R (2015) The role of neutron star mergers in the chemical evolution of the Galactic halo. Astron Astrophys 577:A139
9. Cohen JG, Huang W (2009) The chemical evolution of the draco dwarf spheroidal galaxy. Astrophys J 701:1053–1075

10. Cohen JG, Huang W (2010) The chemical evolution of the ursa minor dwarf spheroidal galaxy. Astrophys J 719:931–949
11. Côté B, Belczynski K, Fryer CL, Ritter C, Paul A, Wehmeyer B, O'Shea BW (2017) Advanced LIGO constraints on neutron star mergers and rprocess sites. Astrophys J 836:230
12. de Boer TJL, Tolstoy E, Hill V, Saha A, Olszewski EW, Mateo M, Starkenburg E, Battaglia G, Walker MG (2012a) The star formation and chemical evolution history of the Fornax dwarf spheroidal galaxy. Astron Astrophys 544:A73
13. de Boer TJL, Tolstoy E, Hill V, Saha A, Olsen K, Starkenburg E, Lemasle B, Irwin MJ, Battaglia G (2012b) The star formation and chemical evolution history of the sculptor dwarf spheroidal galaxy. Astron Astrophys 539:A103
14. Dominik M, Belczynski K, Fryer C, Holz DE, Berti E, Bulik T, Mandel I, O'Shaughnessy R (2012) Double compact objects. I. The significance of the common envelope on merger rates. Astrophys J 759:52
15. Frebel A, Norris JE (2015) Near-field cosmology with extremely metal- poor stars. Ann Rev Astron Astrophys 53:631–688
16. Freiburghaus C, Rosswog S, Thielemann F-K (1999) R-process in neutron star mergers. Astrophys J Lett 525:L121–L124
17. Geisler D, Smith VV, Wallerstein G, Gonzalez G, Charbonnel C (2005) "Sculptor-ing" the Galaxy? The chemical compositions of red giants in the sculptor dwarf spheroidal galaxy. Astron J 129:1428–1442
18. Goriely S, Bauswein A, Janka H-T (2011) r-process nucleosynthesis in dynamically ejected matter of neutron star mergers. Astrophys J Lett 738:L32
19. Hirai Y, Ishimaru Y, Saitoh TR, Fujii MS, Hidaka J, Kajino T (2015) Enrichment of r-process elements in dwarf spheroidal galaxies in chemodynamical evolution model. Astrophys J 814:41
20. Hirai Y, Ishimaru Y, Saitoh TR, Fujii MS, Hidaka J, Kajino T (2017) Early chemo-dynamical evolution of dwarf galaxies deduced from enrichment of r-process elements. Mon Not Royal Astron Soc 466:2474–2487
21. Ishimaru Y, Wanajo S, Prantzos N (2015) Neutron star mergers as the origin of r-process elements in the galactic halo based on the sub-halo clustering scenario. Astrophys J Lett 804:L35
22. Ji AP, Frebel A, Simon JD, Chiti A (2016a) Complete element abundances of nine stars in the r-process galaxy reticulum II. Astrophys J 830:93
23. Ji AP, Frebel A, Chiti A, Simon JD (2016b) R-process enrichment from a single event in an ancient dwarf galaxy. Nature 531:610–613
24. Kinugawa T, Inayoshi K, Hotokezaka K, Nakauchi D, Nakamura T (2014) Possible indirect confirmation of the existence of Pop III massive stars by gravitational wave. Mon Not Royal Astron Soc 442:2963–2992
25. Komiya Y, Shigeyama T (2016) Contribution of neutron star mergers to the r-process chemical evolution in the hierarchical galaxy formation. Astrophys J 830:76
26. Komiya Y, Yamada S, Suda T, Fujimoto MY (2014) The new model of chemical evolution of r-process elements based on the hierarchical galaxy formation. I. Ba and Eu. Astrophys J 783:132
27. Korobkin O, Rosswog S, Arcones A, Winteler C (2012) On the astrophysical robustness of the neutron star merger r-process. Mon Not Royal Astron Soc 426:1940–1949
28. Lemasle B, Hill V, Tolstoy E, Venn KA, Shetrone MD, Irwin MJ, de Boer TJL, Starkenburg E, Salvadori S (2012) VLT/FLAMES spectroscopy of red giant branch stars in the Carina dwarf spheroidal galaxy. Astron Astrophys 538:A100
29. Lorimer DR (2008) Binary and millisecond pulsars. Living Rev Relativ 11:8
30. Mathews GJ, Cowan JJ (1990) New insights into the astrophysical rprocess. Nature 345:491–494
31. Matteucci F, Romano D, Arcones A, Korobkin O, Rosswog S (2014) Europium production: neutron star mergers versus core-collapse supernovae. Mon Not Royal Astron Soc 438:2177–2185
32. Ojima T, Ishimaru Y, Wanajo S, Prantzos N, François P (2018) Stochastic chemical evolution of galactic subhalos and the origin of r-process elements. Astrophys J 865(87):87

33. Roberts LF, Kasen D, Lee WH, Ramirez-Ruiz E (2011) Electromagnetic transients powered by nuclear decay in the tidal tails of coalescing compact binaries. Astrophys J Lett 736:L21

34. Roederer IU, Preston GW, Thompson IB, Shectman SA, Sneden C, Burley GS, Kelson DD (2014) A search for stars of very low metal abundance. VI. Detailed Abundances of 313 metal-poor stars. Astron J 147:136

35. Roederer IU, Mateo M, Bailey JI III, Song Y, Bell EF, Crane JD, Loebman S, Nidever DL, Olszewski EW, Shectman SA, Thompson IB, Valluri M, Walker MG (2016) Detailed chemical abundances in the rprocess-rich Ultra-faint dwarf Galaxy Reticulum 2. Astron J 151:82

36. Rosswog S, Korobkin O, Arcones A, Thielemann F-K, Piran T (2014) The long-term evolution of neutron star merger remnants - I. The impact of r-process nucleosynthesis. Mon Not Royal Astron Soc 439:744–756

37. Ryan SG, Norris JE (1991) Subdwarf studies. III. The halo metallicity distribution. Astron J 101:1865–1878

38. Sadakane K, Arimoto N, Ikuta C, Aoki W, Jablonka P, Tajitsu A (2004) Subaru/HDS abundances in three giant stars in the ursa minor dwarf spheroidal galaxy. Publ Astron Soc Jpn 56:1041–1058

39. Sbordone L, Bonifacio P, Buonanno R, Marconi G, Monaco L, Zaggia S (2007) The exotic chemical composition of the Sagittarius dwarf spheroidal galaxy. Astron Astrophys 465:815–824

40. Shen S, Cooke RJ, Ramirez-Ruiz E, Madau P, Mayer L, Guedes J (2015) The history of R-process enrichment in the milky way. Astrophys J 807:115

41. Shetrone M, Venn KA, Tolstoy E, Primas F, Hill V, Kaufer A (2003) VLT/UVES abundances in four nearby dwarf spheroidal galaxies. I. Nucleosynthesis and abundance ratios. Astron J 125:684–706

42. Shetrone MD, Côté P, Sargent WLW (2001) Abundance patterns in the draco, sextans, and ursa minor dwarf spheroidal galaxies. Astrophys J 548:592–608

43. Shibagaki S, Kajino T, Mathews GJ, Chiba S, Nishimura S, Lorusso G (2016) Relative contributions of the weak, main, and fission-recycling rprocess. Astrophys J 816:79

44. Suda T, Katsuta Y, Yamada S, Suwa T, Ishizuka C, Komiya Y, Sorai K, Aikawa M, Fujimoto MY (2008) Stellar abundances for the galactic archeology (SAGA) satabase—compilation of the characteristics of known exreferences 133 tremely metal-poor stars. Publ Astron Soc Jpn 60:1159–1171

45. Suda T, Yamada S, Katsuta Y, Komiya Y, Ishizuka C, Aoki W, Fujimoto MY (2011) The stellar abundances for galactic archaeology (SAGA) data base - II. Implications for mixing and nucleosynthesis in extremely metal-poor stars and chemical enrichment of the Galaxy. Mon Not Royal Astron Soc 412:843–874

46. Suda T, Hidaka J, Aoki W, Katsuta Y, Yamada S, Fujimoto MY, Ohtani Y, Masuyama M, Noda K, Wada K (2017) Stellar abundances for galactic archaeology database. IV. Compilation of stars in dwarf galaxies. Publ Astron Soc Jpn 69:76

47. Tanaka M et al (2017) Kilonova from post-merger ejecta as an optical and near-Infrared counterpart of GW170817. Publ Astron Soc Jpn 69:102

48. Tsujimoto T, Nishimura N (2015) The r-process in magnetorotational supernovae. Astrophys J Lett 811:L10

49. Tsujimoto T, Shigeyama T (2014) Enrichment history of r-process elements shaped by a merger of neutron star pairs. Astron Astrophys 565:L5

50. van de Voort F, Quataert E, Hopkins PF, Kereš D, Faucher-Giguère C-A (2015) Galactic r-process enrichment by neutron star mergers in cosmological simulations of a Milky Way-mass galaxy. Mon Not Royal Astron Soc 447:140–148

51. Vangioni E, Goriely S, Daigne F, François P, Belczynski K (2016) Cosmic neutron-star merger rate and gravitational waves constrained by the r-process nucleosynthesis. Mon Not Royal Astron Soc 455:17–34

52. Venn KA, Shetrone MD, Irwin MJ, Hill V, Jablonka P, Tolstoy E, Lemasle B, Divell M, Starkenburg E, Letarte B, Baldner C, Battaglia G, Helmi A, Kaufer A, Primas F (2012) Nucleosynthesis and the inhomogeneous chemical evolution of the carina dwarf galaxy. Astrophys J 751:102

53. Wanajo S, Sekiguchi Y, Nishimura N, Kiuchi K, Kyutoku K, Shibata M (2014) Production of
    All the r-process nuclides in the dynamical ejecta of neutron star mergers. Astrophys J Lett
    789:L39
54. Wehmeyer B, Pignatari M, Thielemann F-K (2015) Galactic evolution of rapid neutron capture
    process abundances: the inhomogeneous approach. Mon Not Royal Astron Soc 452:1970–1981
55. Yamada S, Suda T, Komiya Y, Aoki W, Fujimoto MY (2013) The stellar abundances for
    galactic archaeology (SAGA) database - III. Analysis of enrichment histories for elements
    and two modes of star formation during the early evolution of the MilkyWay. Mon Not Royal
    Astron Soc 436:1362–1380

Chapter 6
Efficiency of Metal Mixing in Dwarf Galaxies

Abstract Stars inherit the elemental abundances of ejecta from nucleosynthetic events mixed in the interstellar medium. The abundances of heavy elements in stars contain information about the metal mixing in galaxies. However, there are a few constraints about the efficiency of metal mixing in galaxies. In this chapter, we propose that the abundances of heavy elements can be used as an indicator to understand the efficiency of metal mixing. We show that the efficiency of metal mixing is difficult to constrain by only using α-element abundances. On the other hand, the r-process elements can be a good indicator to constrain the efficiency of metal mixing. We find that the scaling factor for metal mixing should be larger than 0.01 to explain the observed low abundances of r-process elements in dwarf spheroidal galaxies. This value corresponding to the timescale of metal mixing less than 40 million years, which is shorter than the dynamical times of dwarf galaxies (Contents in this chapter have been in part published in Hirai et al. [6–8] reproduced by permission of the AAS).

6.1 Heavy Elements as an Indicator of Metal Mixing

Metal mixing is a highly complex phenomenon occurring in a different scales from the region to the star formation ($\lesssim 10$ pc) to the scale in the cluster of galaxies (~ 1 Mpc). Metal mixing is mainly caused by interstellar turbulence. It is powered by different sources such as galactic interactions, galactic rotation, HII regions, and supernovae (e.g., [2]). Power-law slopes of the power spectra of HI emission, absorption, and CO emission can be observed to constrain the nature of the interstellar turbulence.

Metal mixing in galaxies plays an important role to determine abundances of stars (e.g., [13]). Signatures of metal mixing can be obtained from the abundances of elements in metal-poor stars. There are star-to-star scatters less than 0.2 dex in the ratios of α-elements to iron (Fig. 1.8). On the other hand, there is an increasing trend toward lower metallicity with small scatters in [Zn/Fe] ratios (Fig. 1.9).

© Springer Nature Singapore Pte Ltd. 2019
Y. Hirai, *Understanding the Enrichment of Heavy Elements
by the Chemodynamical Evolution Models of Dwarf Galaxies*,
Springer Theses, https://doi.org/10.1007/978-981-13-7884-3_6

The most characteristic feature in the abundance ratios can be seen in the r-process elements. At [Fe/H] $\lesssim -2.5$, there are no stars with [Ba/Fe] $\gtrsim 1$ in the Local Group dwarf spheroidal galaxies while several r-process enhanced stars in the Milky Way halo and the ultra-faint dwarf galaxy Reticulum II (e.g., [4, 12, 18]). These observational features could be used to constrain the efficiency of metal mixing in galaxies.

Astrophysical sites of elements affect the distribution of [elements/Fe] ratios. Core-collapse supernovae synthesize α-elements as well as iron (e.g., [15]). In the case of α-elements, the scatters of [α-elements/Fe] are intrinsically small. The small scatters of α-elements seen in the observation reflect this fact. On the other hand, large scatters of r-process elements indicate that not all core-collapse supernovae synthesize r-process elements. Chapter 5 shows that neutron star mergers can be the astrophysical sites of r-process elements. These results imply that the [r/Fe] ratios can easily reflect the inhomogeneity in the distribution of elements.

Hydrodynamic simulations can constrain the metal mixing in galaxies. Shen et al. [20] suggested that metal content in the intergalactic medium decreases due to metal diffusion. Revaz et al. [17] studied the scatters of [α/Fe] ratios with different strength of metal diffusion. Their results suggested that the scatters of [α/Fe] ratios are too large to explain the observations without assuming the effects of metal mixing. Escala et al. [3] have shown that it is necessary to include metal diffusion models in SPH simulations to reproduce metallicity distributions of dwarf galaxies.

The efficiency and timescales of metal mixing are not well clarified. It is difficult to estimate the efficiency of metal mixing by only considering [α/Fe] ratios due to their small scatters. Revaz et al. [17] showed that the optimal value for metal diffusion constrained in simulations of dwarf galaxies is not enough to reproduce the [α/Fe] ratios in the Milky Way-like galaxies. These results indicate that it is necessary to explore another indicator of metal mixing.

This chapter aims to explore the metal mixing in dwarf galaxies using the elemental abundances in metal-poor stars. We adopt Mg, Zn, and Ba as indicators of metal mixing. In this chapter, an isolated dwarf galaxy model is adopted (see Chap. 3 for details). The model has the same parameters with model G in Chap. 4. Table 6.1 lists adopted values of C_d in this chapter. Section 6.2 compares results with a different efficiency of metal mixing. Section 6.3 shows the problems arising from the SPH simulations without assuming metal mixing models. Section 6.4 discuss the efficiency of metal mixing implied from the Ba abundances.

Table 6.1 The scaling factors for metal diffusion coefficient adopted in this chapter

Model	C_d
d1000	10^{-1}
d0100	10^{-2}
d0010	10^{-3}
d0001	10^{-4}
d0000	

6.2 Effects of Metal Mixing on the Enrichment of Mg and Zn

The ratios of [Mg/Fe] cannot constrain the efficiency of metal mixing but show the necessity of metal mixing models in SPH simulations. Figure 6.1 represents [Mg/Fe] as a function of [Fe/H]. According to this figure, model d000 (Fig. 6.1e) shows scatters of [Mg/Fe] ratios, which are not seen in the observations (red hatched area in Fig. 6.1f). On the other hand, the [Mg/Fe] ratios in models with metal mixing lie in the observed scatters at [Fe/H] < −2 (Fig. 6.1a–d). This result suggests that it is necessary to include metal mixing in SPH simulations.

The ratios of [Zn/Fe] can also be an indicator for metal mixing. Figure 6.2 denotes [Zn/Fe] as functions of [Fe/H] predicted by models d0010 and d1000. As shown in this figure, model d0010 (Fig. 6.2a) has larger scatters of [Zn/Fe] ratios than those in model d1000 (Fig. 6.2b). The number of stars which have [Zn/Fe] > 0.5 is larger in model d0010 than that in model d1000. This is because the model d1000 more efficiently mixes metals than model d0010.

These results suggest that we need to implement models of metal mixing in SPH simulations of galaxies. Abundances of elements are locked in gas particles in SPH simulations that do not assume metal mixing models. Lack of metal mixing causes artificial scatters in the abundances of elements. However, it is difficult to constrain the efficiency of metal mixing from the ratios of [Mg/Fe] and [Zn/Fe]. From the next section, we show the efficiency of metal mixing can be constrained by using ratios of r-process elements to iron.

6.3 Effects of the Metal Mixing on the Enrichment of r-Process Elements

Enrichment of r-process elements in SPH simulations without metal mixing causes a severe problem. Figure 6.3 shows [Eu/Fe] as functions of [Fe/H] without metal mixing models. As shown in this figure, there are large star-to-star scatters of [Eu/Fe]. There are stars of [Eu/Fe] < 0, which are not seen in the observations. This is due to the inhomogeneity caused by the fact that SPH particles cannot diffuse elements to the other particles.

Figure 6.4a denotes time evolution of [Fe/H]. As shown in this figure, the dispersion of [Fe/H] over 1 dex can be seen even at 5 billion years from the beginning of the simulation. Black plots in this figure represent stars which inherit the ejecta from one core-collapse supernova. According to this figure, these stars have over 3 dex dispersion in the [Fe/H] ratios.

Figure 6.4b depicts time evolution of [Eu/H]. This figure shows that star-to-star scatters of [Eu/H] are not erased throughout the evolutionary history of the galaxy. According to this figure, stars which inherit Eu from only one neutron star merger

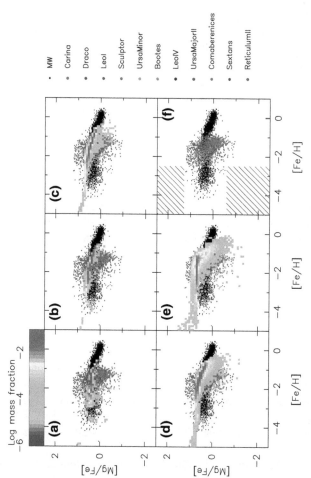

Fig. 6.1 [Mg/Fe] as functions of [Fe/H] in stars (Fig. 1 of [6] reproduced by permission of the AAS). We plot models d1000, d0100, d0010, d0001, and d0000 on panels (**a**), (**b**), (**c**), (**d**), and (**e**), respectively. We only plot observational data in panel (**f**). The region where there is no observation of [Mg/Fe] is shown with the red hatched area. The color-coded region represents the logarithm of mass fractions of computed stars. Black dots denotes the abundances of the Milky Way halo stars. The color dots show the abundances on the stars in the Local Group dwarf galaxies: Carina (green), Draco (green), Leo I (purple), Sculptor (magenta), Ursa Minor (ocher), Boötes I (orange), Leo IV (dark green), Ursa Major II (grass green), Coma Berenices (sky-blue), Sextans (red-purple), and Reticulum II (red). We compile all observed data using the SAGA database [21–23, 26]

Fig. 6.2 Same as Fig. 4.1, but for models d0010 and d1000

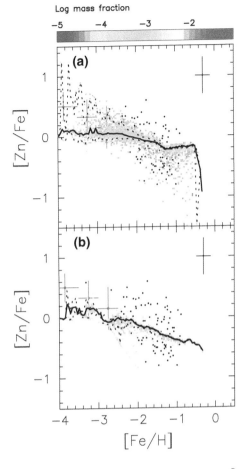

Fig. 6.3 Same as Fig. 5.1, but for model s000 (Fig. 5 of [7], reproduced by permission of the AAS)

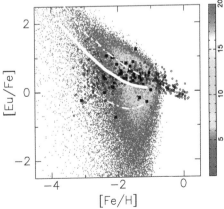

Fig. 6.4 Time evolution of **a**
[Fe/H] and **b** [Eu/H] in
model s000. Black circles
represents stars formed from
gas particles which inherit
one **a** core-collapse
supernova and **b** neutron star
merger. The number of
computed stars are described
in color from 0 (purple) to 20
(red). (Fig. 6 of [7]
reproduced by permission of
the AAS)

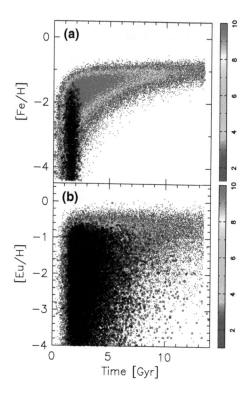

can be formed over 14 billion years. This is due to the low rates of neutron star
mergers and insufficient metal mixing.

Metal mixing reduces the fraction of stars with extreme r-process abundances.
Figure 6.5 compares the [Eu/Fe] ratios of stars with [Fe/H] < -2.0 in models without
(left panel) and with (right panel) metal mixing in star-forming region. In the model
without metal mixing, dispersion of [Eu/Fe] ratios and the fraction of r-process
enhanced stars are too large compared to the observations. As shown in this figure,
the dispersion of [Eu/Fe] ratios significantly reduced if we implement the metal
mixing.

The efficiency of metal mixing may depend on the size of the metal mixing region
and resolution. Figure 6.6a shows effects of the adopted number of nearest neighbor
particles, which directly obtains the ejecta of supernovae or neutron star mergers
when they explode (N_{ngb}). If we adopt the larger value of N_{ngb}, scatters of [Eu/Fe]
become smaller. Here we define the metal mixing mass as $M_{mix} = N_{ngb} \times m_{gas}$.
Table 6.2 lists this value. When we increase the number of neighbor particles, the
value of M_{mix} also increases.

In contrast, the mass resolution does not affect the ratios of [Eu/Fe] in this model.
According to Fig. 6.6b, scatters of models m017, m018, and m019 are overlapped.

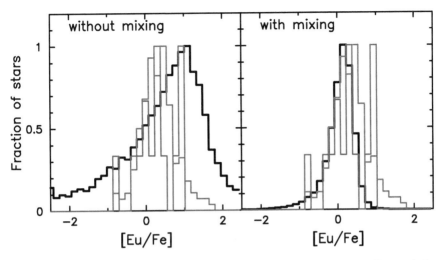

Fig. 6.5 Distribution of [Eu/Fe] in stars with [Fe/H] < -2.0 (Fig. 8 of [7] reproduced by permission of the AAS). Black histograms show the results of model s000 (left panel) and m000 (right panel). Red and blue histograms represent the observed values in the Milky Way halo and the Local Group dwarf galaxies (Carina, Draco, Leo I, Sculptor, and Ursa Minor), respectively

Table 6.2 The mass of metals mixed in star forming region

Model	N	N_{ngb}	M_{mix} $(10^4 M_\odot)$
m000	2^{19}	32	1.3
mN16	2^{19}	16	0.6
mN64	2^{19}	64	2.7
m018	2^{18}	32	2.6
m017	2^{17}	32	5.1
m016	2^{16}	32	10.3
m014	2^{14}	32	41.0

This is because the value of the mixing mass in a star-forming region is proportional to the gas mass. Gas mass is also proportional to the inverse square of the number of particles. On the other hand, the number of star-forming regions, which mixes the metals is proportional to the number of particles. These assumptions offset the effects of the mass resolution of the simulations.

More detailed models of metal mixing enable us to constrain the efficiency of metal mixing. Here we compare the results computed with turbulence-motivated metal diffusion models. Figure 6.7 shows the *r*-process components of [Ba/Fe] as a function of [Fe/H] with different efficiencies of metal mixing. According to this figure, the computed [Ba/Fe] ratios highly depend on the efficiencies of metal mixing. Stars with [Ba/Fe] > 1 are formed in models with the scaling factor of metal diffusion (C_d) lower than 10^{-3} (Fig. 6.7 c–e). These stars are not seen in the observations. On

Fig. 6.6 Effects of **a** the number of nearest neighbor particles and **b** the initial number of particles on [Eu/Fe] as a function of [Fe/H] (Fig. 9 of [7] reproduced by permission of the AAS). Solid and dashed curves show the median, first and third quartiles, respectively. Red, black, green, and blue curves in panel **a** represent models mN16, m000, mN64, and s000, respectively. Red, green, blue, sky blue, black curves in panel **b** represent models m014, m016, m017, m018, and m000

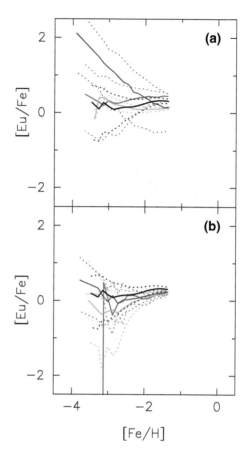

the other hand, models with $C_d \gtrsim 10^{-2}$ do not have such r-process enhanced stars (Fig. 6.7 a, b). This result is consistent with the observations of the Local Group dwarf spheroidal galaxies.

The scaling factor for metal diffusion, C_d controls the inhomogeneity of gas phase metallicity. Models with higher values of C_d quickly erases the gases with metallicity which profoundly deviates from the average value. As a result, a formation of stars with extreme abundances are suppressed in these models. When neutron star mergers occur in a galaxy, gases around them temporary have [Ba/Fe] $\gtrsim 1$. Elements distributed in gas diffuse into the surrounding gases. The r-process enhanced stars can only be formed when the timescale of metal mixing is longer than the timescale of star formation in a galaxy.

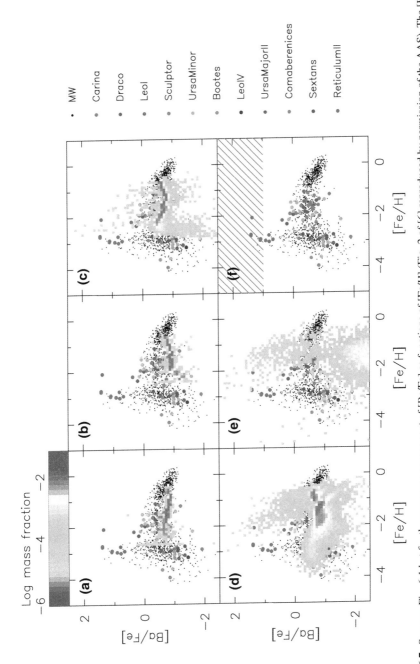

Fig. 6.7 Same as Fig. 6.1 but for the *r*-process component of [Ba/Fe] as functions of [Fe/H] (Fig. 2 of [6] reproduced by permission of the AAS). The [Ba/Fe] ratios in all stars with [Fe/H] > −2.75 are corrected to be [Ba/Eu] = −0.89 to eliminate the *s*-process contribution

6.4   Efficiency of Metal Mixing

Results of the previous section suggest that the scaling factor for metal diffusion should be higher than 10^{-2} when we adopt the shear-based metal diffusion model. Scaling factors for metal diffusion can be interpreted as a timescale of metal mixing (t_{mix}). In this study, we define t_{mix} as an average time required to reduce gas phase [Ba/Fe] ratios from [Ba/Fe] > 1 to [Ba/Fe] < 0. In models d1000 and d0100, the values of t_{mix} are $\lesssim 10$ and $\simeq 40$ million years, respectively. Note that it is not possible to estimate the timescale less than 10 million years because this simulation produces a snapshot with the time interval of 10 million years. On the other hand, the values of t_{mix} in models d0010 and d0001 are $\simeq 360$ million years and 1.6 billion years, respectively. These results suggest that metals should be mixed with shorter timescales than the dynamical times of dwarf galaxies (typically ~ 100 million years).

The diffusion coefficient constrained in this study is consistent with the values estimated from theories and experiments in smaller scales. The value of C_d is 0.0324 when we adopt a Kolmogorov constant as 1.41 (e.g., [14]). The optimal values of C_d in a turbulent channel flow and a turbulent mixing layer are 0.01 and 0.0225, respectively [9]. Although the Smagorinsky model adopted in this study depends on the flow field, our results imply that we can test the Smagorinsky model on a scale of galaxies.

The efficiency of metal mixing corresponding to $C_d \gtrsim 0.01$ is also consistent with Ji et al. [11]. They found that Population II (metal-poor) supernovae efficiently wipe out the signatures from Population III (zero metal) stars. Diffusion coefficients adopted in their study are 2.4×10^{-5} kpc^2 Myr^{-1} (mini-halo models) and 8.1×10^{-4} kpc^2 Myr^{-1} (atomic cooling halo models), respectively. The average value of diffusion coefficient at one billion years from the start of the simulations in this study is 2×10^{-4} kpc^2 Myr^{-1} (model d1000) and 2×10^{-5} kpc^2 Myr^{-1} (model d0100), respectively. From this result, we confirm that it is possible to draw similar conclusions with Ji et al. [11].

The results are not significantly affected by adopted diffusion models. Here we adopt the shear-based metal diffusion model [19, 20]. On the other hand, Greif et al. [5] constructed a velocity dispersion based metal diffusion model. Williamson et al. [25] have shown that diffusion coefficient computed with the shear-based model [20] is about half as large as that of the velocity dispersion based model [5]. Williamson et al. [25] also found that the models for diffusion do not affect the properties of gases. We confirm that the results are not significantly altered when we adopt the model of Greif et al. [5].

We need to increase the number of observations of metal-poor stars in the Local Group dwarf galaxies to constrain the efficiency of metal mixing more precisely. It is necessary to determine the degree of scatters in the abundances of r-process elements within each galaxy. Recent observations add samples of metal-poor stars (e.g., [1, 24]). Metallicity distribution functions may also be an indicator for metal mixing (e.g., [3, 16]). We also find that models with more efficient metal mixing produce the smaller fraction of extremely metal-poor stars. We expect that the efficiency of metal mixing in galaxies can be constrained more precisely by comparing the results of future simulations and observations.

The initial conditions of the simulations may affect the abundances of the lowest metallicity stars. Jeon et al. [10] have shown that stars with the lowest metallicity are formed in accreted halos polluted by gases outside of the halo. If we consider cosmological initial conditions, the lowest metallicity stars can be formed by such material. In the next chapter, we discuss the enrichment of r-process elements in a cosmological context.

6.5 Summary

In this chapter, we performed a series of N-body/SPH simulations of dwarf galaxies with a turbulence motivated metal diffusion model. We find that the scaling factor for metal diffusion larger than 0.01 is required to reproduce the observation of abundances of heavy elements in dwarf galaxies. We confirm that turbulence theory and experiment also estimate the similar value. Also, we find that the timescale need to erase gases with an enhanced r-process abundances is $\lesssim 40$ million years, which is shorter than the dynamical times of the system. These results suggest that the scaling factor for metal diffusion of 0.01 appears to be suitable for smoothed particle hydrodynamics simulations of galaxies with the shear-based metal-diffusion model.

References

1. Duggan GE, Kirby EN, Andrievsky SM, Korotin SA (2018) Neutron star mergers are the dominant source of the r-process in the early evolution of dwarf galaxies. Astrophys J 869(50):50
2. Elmegreen BG, Scalo J (2004) Interstellar turbulence I: observations and processes. Ann Rev Astron Astrophys 42:211–273
3. Escala I, Wetzel A, Kirby EN, Hopkins PF, Ma X, Wheeler C, Kereš D, Faucher-Giguére C-A, Quataert E (2018) Modelling chemical abundance distributions for dwarf galaxies in the Local Group: the impact of turbulent metal diffusion. Monthly Notices of the Royal Astronomical Society 474:2194–2211
4. Frebel A, Norris JE (2015) Near-field cosmology with extremely metal-poor stars. Ann Rev Astron Astrophys 53:631–688
5. Greif TH, Glover SCO, Bromm V, Klessen RS (2009) Chemical mixing in smoothed particle hydrodynamics simulations. Mon Not Royal Astron Soc 392:1381–1387
6. Hirai Y, Saitoh TR (2017) Efficiency of metal mixing in dwarf galaxies. Astrophys J Lett 838:L23
7. Hirai Y, Ishimaru Y, Saitoh TR, Fujii MS, Hidaka J, Kajino T (2015) Enrichment of r-process elements in dwarf spheroidal galaxies in chemodynamical evolution model. Astrophys J 814:41
8. Hirai Y, Saitoh TR, Ishimaru Y, Wanajo S (2018) Enrichment of Zinc in galactic chemodynamical evolution models. Astrophys J 855(63):63
9. Horiuti K (1987) Comparison of conservative and rotational forms in large eddy simulation of turbulent channel flow. J Computat Phys 71:343–370
10. Jeon M, Besla G, Bromm V (2017) Connecting the first galaxies with ultrafaint dwarfs in the Local Group: chemical signatures of population III stars. Astrophys J 848(85):85

11. Ji AP, Frebel A, Bromm V (2015) Preserving chemical signatures of primordial star formation in the first low-mass stars. Mon Not Royal Astron Soc 454:659–674
12. Ji AP, Frebel A, Chiti A, Simon JD (2016) R-process enrichment from a single event in an ancient dwarf galaxy. Nature 531:610–613
13. Karlsson T, Bromm V, Bland-Hawthorn J (2013) Pregalactic metal enrichment: the chemical signatures of the first stars. Rev Modern Phys 85:809–848
14. Meneveau C, Katz J (2000) Scale-invariance and turbulence models for large-eddy simulation. Ann Rev Fluid Mech 32:1–32
15. Nomoto K, Kobayashi C, Tominaga N (2013) Nucleosynthesis in stars and the chemical enrichment of galaxies. Ann Rev Astron Astrophys 51:457–509
16. Pilkington K, Gibson BK, Brook CB, Calura F, Stinson GS, Thacker RJ, Michel-Dansac L, Bailin J, Couchman HMP, Wadsley J, Quinn TR, Maccio A (2012) The distribution of metals in cosmological hydrodynamical simulations of dwarf disc galaxies. Mon Not Royal Astron Soc 425:969–978
17. Revaz Y, Arnaudon A, Nichols M, Bonvin V, Jablonka P (2016) Computational issues in chemo-dynamical modelling of the formation and evolution of galaxies. Astron Astrophys 588:A21
18. Roederer IU, Mateo M, Bailey JI III, Song Y, Bell EF, Crane JD, Loebman S, Nidever DL, Olszewski EW, Shectman SA, Thompson IB, Valluri M, Walker MG (2016) Detailed chemical abundances in the rprocess- rich ultra-faint dwarf galaxy reticulum 2. Astron J 151:82
19. Saitoh TR (2017) Chemical evolution library for galaxy formation simulation. Astronom J 153:85
20. Shen S, Wadsley J, Stinson G (2010) The enrichment of the intergalactic medium with adiabatic feedback - I. Metal cooling and metal diffusion. Mon Not Royal Astron Soc 407:1581–1596
21. Suda T, Katsuta Y, Yamada S, Suwa T, Ishizuka C, Komiya Y, Sorai K, Aikawa M, Fujimoto MY (2008) Stellar abundances for the galactic archeology (SAGA) database—compilation of the characteristics of known extremely metal-poor stars. Publ Astron Soc Jpn 60:1159–1171
22. Suda T, Yamada S, Katsuta Y, Komiya Y, Ishizuka C, Aoki W, Fujimoto MY (2011) The stellar abundances for galactic archaeology (SAGA) data base - II. Implications for mixing and nucleosynthesis in extremely metal-poor stars and chemical enrichment of the Galaxy. Mon Not Royal Astron Soc 412:843–874
23. Suda T, Hidaka J, Aoki W, Katsuta Y, Yamada S, Fujimoto MY, Ohtani Y, Masuyama M, Noda K, Wada K (2017) Stellar abundances for galactic archaeology database. IV. Compilation of stars in dwarf galaxies. Publications of the Astronomical Society of Japan 69, 76
24. Tsujimoto T, Matsuno T, Aoki W, Ishigaki MN, Shigeyama T (2017) Enrichment in r-process elements from multiple distinct events in the early draco dwarf spheroidal galaxy. Astrophys J 850(L12):L12
25. Williamson D, Martel H, Kawata D (2016) Metal diffusion in smoothed particle hydrodynamics simulations of dwarf galaxies. Astrophys J 822:91
26. Yamada S, Suda T, Komiya Y, Aoki W, Fujimoto MY (2013) The stellar abundances for galactic archaeology (SAGA) database - III. Analysis of enrichment histories for elements and two modes of star formation during the early evolution of the MilkyWay. Mon Not Royal Astron Soc 436:1362–1380

Chapter 7
Enrichment of *r*-Process Elements in a Cosmological Context

Abstract This chapter connects the enrichment of *r*-process elements to the formation of Local Group galaxies. Enrichment of *r*-process elements is associated with galaxy formation. In the hierarchical structure formation scenario, galaxies are thought to be formed by clustering of smaller systems. Star-to-star scatters of the ratios of *r*-process elements in the Milky Way halo should be reflected such process. However, the enrichment of *r*-process elements in a cosmological context is not yet understood. Here we performed a series of high-resolution cosmological zoom-in simulations of galaxies with halo mass of $10^{10} \, M_\odot$ at redshift 2. We find that most of the extremely metal-poor stars are formed before 1 billion years from the beginning of the simulation. In the simulations, there are three *r*-process rich extremely metal-poor stars. All of these stars are formed in the halo with a gas mass of $10^6 \, M_\odot$ at the early stages of galaxy formation. This result suggests that *r*-process rich stars seen in the Milky Way halo come from small size halos with the mass comparable to the Local Group ultra-faint dwarf galaxies.

7.1 Enrichment of *r*-Process Elements and Galaxy Formation

Abundances of *r*-process elements in metal-poor stars help us understand the early stages of the Milky Way halo formation. Astronomical high-dispersion spectroscopic observations have shown that there are star-to-star scatters of over 3 dex in the Milky Way halo at [Fe/H] ≤ -2.5 (e.g., [7]). Beers and Christlieb [4] classified *r*-process rich stars. According to their classification, stars with $0.3 \leq$ [Eu/Fe] ≤ 1.0 and [Ba/Eu] < 0 are called r-I stars. On the other hand, stars with [Eu/Fe] > 1.0 and [Ba/Eu] < 0 are r-II stars. As shown in Fig. 1.10, r-II stars have not found in the Local Group dwarf spheroidal galaxies. Ji et al. [15] discovered seven r-II stars in Reticulum II ultra-faint dwarf galaxy. Roederer et al. [23] also found two r-II stars in their sample of Reticulum II. Another ultra-faint dwarf galaxies have depleted abundances of *r*-process elements (e.g., [7]).

© Springer Nature Singapore Pte Ltd. 2019
Y. Hirai, *Understanding the Enrichment of Heavy Elements by the Chemodynamical Evolution Models of Dwarf Galaxies*,
Springer Theses, https://doi.org/10.1007/978-981-13-7884-3_7

The most promising astrophysical sites of r-process elements are neutron star mergers. Nucleosynthetic studies have shown that neutron star mergers can synthesize r-process elements heavier than $A = 110$ (e.g., [3, 8, 10, 18, 26, 35]). The existence of neutron star mergers is confirmed by the detection of gravitational waves and multi-messenger observations of a neutron star merger, GW170817 (e.g., [1, 2]). The optical and infrared afterglow of GW170817 suggested that they produced $\approx$ 0.03 $M_\odot$ of r-process elements (e.g., [32, 33]).

Chemical evolution studies taking into account the hierarchical structure formation scenario successfully explained the observed r-process abundances in the Milky Way by neutron star mergers. Ishimaru et al. [14] found that the [Eu/Fe] ratio increases at lower metallicity in halos with smaller star formation efficiencies in their one-zone chemical evolution model. In Chap. 5, we have shown that r-process abundances in these halos are consistent with observation because their star formation rates are suppressed by supernova feedback, resulting in slow chemical evolution. Komiya and Shigeyama [17] showed that the ejecta from neutron star mergers can enrich intergalactic medium by their semi-analytic models.

Previous cosmological hydrodynamic simulations also showed that there are good agreements between simulated results and observations of r-process abundances in the Milky Way halo. Shen et al. [30] reported that their results are insensitive to the merger rate and delay time distributions. van de Voort et al. [34] showed that metal-poor stars with r-process elements are formed in high redshift and large galactocentric radius. Naiman et al. [19] suggested that the Eu abundance does not correlate with assembly histories and properties of galaxies. However, their resolution is not enough to resolve the scale of satellite dwarf galaxies. Ojima et al. [20] showed that r-II stars would come from accreted halos with the mass similar to observed ultra-faint dwarf galaxies based on the model of Ishimaru et al. [14]. It thus needs to perform simulations that can resolve the scale of satellite dwarf galaxies.

In this chapter, we aim to discuss the enrichment of r-process elements in a cosmological context with high-resolution cosmological zoom-in simulations. In Sect. 7.2, we describe our code and cosmological zoom-in simulations. In Sect. 7.3, we show the results of our simulations. In Sect. 7.4, we discuss the formation site of r-II stars.

7.2 Cosmological Zoom-In Simulations

We have performed a series of cosmological "zoom-in" simulations. A preflight low-resolution cosmological N-body simulation is performed with GADGET- 2 [31]. We adopt the ΛCDM cosmology with a total matter density $\Omega_m = 0.308$, a dark energy density $\Omega_\Lambda = 0.692$, a baryon density $\Omega_b = 0.0484$, a Hubble constant $H_0 = 67.8\,\mathrm{km\ s^{-1}}$, an amplitude of the matter power spectrum $\sigma_8 = 0.815\,\mathrm{km\ s^{-1}}$, and a tilted scalar spectral index $n_s = 0.968$ [22]. The initial condition is generated by MUSIC using second-order Lagrangian perturbation theory [11]. The box size is $(4.0\,h^{-1}\,\mathrm{Mpc})^3$ with 256^3 particles. Mass of one dark matter particle is $3.3 \times 10^5\,M_\odot$.

The gravitational softening length is $625\,h^{-1}$ pc. The initial redshift of this simulation is $z_i = 100$.

We then select the halo-of-interest from the snapshot at $z = 0$ of the preflight simulation. For halo find, we use the AMIGA Halo Finder (AHF; [9, 16]). We have selected a halo with a total mass of $2.72 \times 10^{10}\ M_\odot$ at $z = 0$. This halo mass is smaller than that of the Milky Way and larger than those of dwarf spheroidal galaxies. Ideally, we should compute the Milky Way-mass halos. However, it is difficult to compute such halos with sufficient resolution to resolve satellite dwarf galaxies. We thus chose the halo with $10^{10}\ M_\odot$ with small box size. The zoomed-in initial condition is generated by MUSIC. The cuboid Lagrangian volume for re-simulation is carefully selected to avoid contamination of low-resolution particles.

After selecting the halos, we performed a series of cosmological zoom-in simulations using ASURA [28, 29] described in Sect. 2.1. A symmetrized form of the softened gravitational potential is adopted to deal with systems which have particles with different gravitational softening lengths [27]. The effective resolutions of zoomed regions are 512^3 particles. The total number of particles in the zoomed-in region is 2.51×10^6. We set gravitational softening lengths for gas and dark matter particles 20 and 50 pc, respectively. Masses of one gas and dark matter particles are $6.4 \times 10^3\ M_\odot$ and $3.4 \times 10^4\ M_\odot$, respectively.

7.3 Enrichment of r-Process Elements in Cosmological Simulations

In this section, we show the result at $z = 2.0$ because this halo experiences a major merger at $z = 1.0$, which is not thought to occur in the Milky Way halo. We confirm that most of the extremely metal-poor stars are formed at $z > 2.0$ in this simulation. The central halo mass at $z = 2.0$ is $M_{tot} = 1.4 \times 10^{10}\ M_\odot$. Throughout this chapter, we define M_{tot} as the virial mass of halos.

Figure 7.1 shows stellar and gas surface density at $z = 2.0$. As shown in this figure, a spiral galaxy is formed in this simulation. The stellar mass and gas mass are $1.4 \times 10^9\ M_\odot$ and $3.3 \times 10^8\ M_\odot$, respectively. The stellar mass to halo mass ratio is $M_*/M_{tot} = 0.1$. This value is larger than that estimated from the abundance matching ($M_*/M_{tot} = 0.002$; [5]). This is because we only assume a thermal form of supernova feedback. This result suggests that another form of feedbacks such as radiation pressure from massive stars [13] and kinetic form of supernova feedback [21] to reproduce the stellar to halo mass ratio.

Figure 7.2 shows star formation rates of the central galaxy within 30 kpc from the galactic center. The star formation rate at $t < 0.6$ billion years is dominated by metal-poor stars with [Fe/H] ≤ -1. According to this figure, most extremely metal-poor stars ([Fe/H] ≤ -3, orange curve) are formed at $t < 1.0$ billion years from the beginning of the simulation. A peak of the star formation rate at 1.67 billion years is due to a merger of a galaxy. Figure 7.3 shows the metallicity distribution of the halo

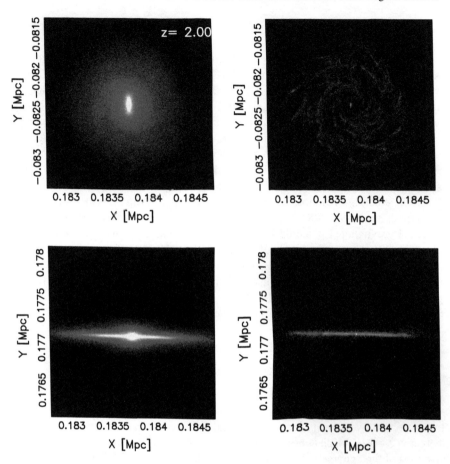

Fig. 7.1 Stellar and gas surface density at $z = 2.0$. Top left panel shows face-on view of stellar surface density. Top right panel shows face-on view of gas surface density. Bottom left panel shows edge-on view of stellar surface density. Bottom right panel shows edge-on view of gas surface density. From black to while, the color scale denotes surface density from $10^{1.0}$ to $10^{3.8}$ $M_\odot \mathrm{pc}^{-2}$

of the simulated galaxy. The median [Fe/H] is [Fe/H] $= -1.70$. This value is close to the peak value ([Fe/H] ≈ -1.7) of metallicity distribution of the Milky Way halo. Figure 7.4 shows [Mg/Fe] as a function of [Fe/H]. The decrease of [Mg/Fe] is due to the contribution of type Ia supernovae. Scatters of [Mg/Fe] are ~ 1 dex at [Fe/H] $= -3.0$. Low [Mg/Fe] stars around [Fe/H] $= -2$ are due to accreted dwarf galaxies.

Figure 7.5 shows [Eu/Fe] as a function of [Fe/H] within 30 kpc from the center of the galaxy. At [Fe/H] > -1.5, most stars located around [Eu/Fe] ~ 0.0. These stars are in the bulge and disk of the galaxy. The small scatter (~ 1 dex) at

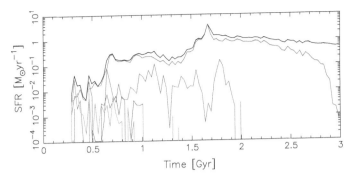

Fig. 7.2 Time variations of star formation rates. The star formation rates are plotted within 30 kpc from the center of the galaxy. The black curve denotes the total star formation rate. Colored curves represent formation rates of stars with different metallicities. Red, green, blue, and orange curves denote formation rates of stars with $-1 < $ [Fe/H] $\leq 0, -2 < $ [Fe/H] $\leq -1, -3 < $ [Fe/H] ≤ -2, and [Fe/H] ≤ -3, respectively

Fig. 7.3 Metallicity distribution of the halo. We plot the metallicity distribution within 30 kpc from the galactic center and 1 kpc above the galactic disk

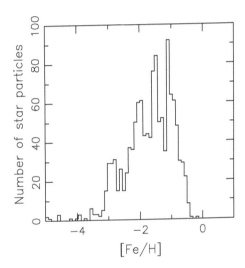

[Fe/H] > -1.5 suggests that spatial distribution of metallicity is homogeneous at the site of the formation of the bulge and disk. On the other hand, at [Fe/H] < -1.5, there are star-to-star scatters of over 3 dex in [Eu/Fe] ratios. Most of these stars are in the halo.

7.4   The Origin of *r*-Process Enhanced Stars

Here we discuss the formation sites of r-II stars. In Chap. 6, we show that metals should be mixed within $\sim$40 million years to explain the observed low abundances

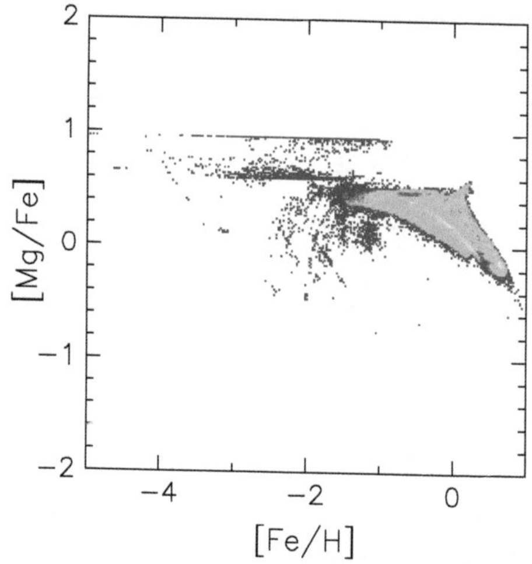

Fig. 7.4 [Mg/Fe] ratios as a function of [Fe/H] within 30 kpc from the center of the galaxy. The color scale is the same as in Fig. 4.1

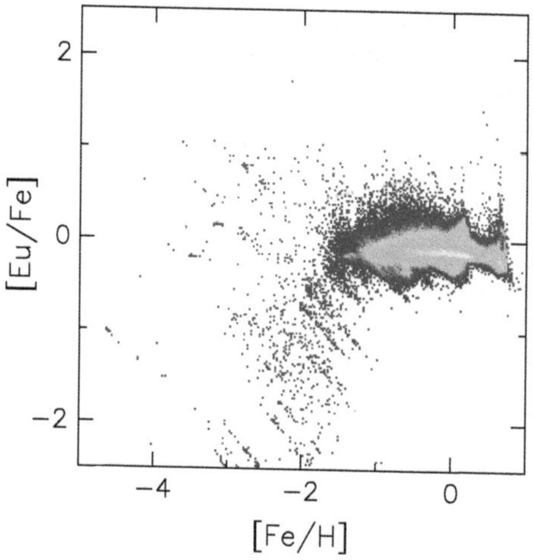

Fig. 7.5 [Eu/Fe] ratios as a function of [Fe/H] within 30 kpc from the center of the galaxy. The color scale is the same as in Fig. 4.1

of *r*-process elements in the Local Group dwarf spheroidal galaxies. On the other hand, [15, 23] reported that there are enhanced *r*-process abundances in Reticulum II ultra-faint dwarf galaxy. The Milky Way halo also has several r-II stars (e.g., [7, 12]).

Figure 7.5 shows there are r-II stars in [Fe/H] < −2 even if we adopt the same scaling factor for metal diffusion with Chap. 6. We find three *r*-process enhanced stars at [Fe/H] < −3 in our simulations. Two stars are formed at 0.37 billion years from the beginning of the simulation. These stars are formed in a halo with $M_{tot} = 1.32 \times 10^8 \, M_\odot$ and $M_{gas} = 8.59 \times 10^6 \, M_\odot$. The other star is formed at 0.45 billion years in a halo with $M_{tot} = 1.32 \times 10^8 \, M_\odot$ and $M_{gas} = 7.08 \times 10^6 \, M_\odot$. Although these stars are formed in different halos, the gas mass of halos that formed r-II stars is similar.

The enhancement of *r*-process abundances can be understood regarding the mass of halos that formed stars with *r*-process elements. Here we assume that one neutron star merger distributes 10^{-2} $M_\odot$ of *r*-process elements and these elements are mixed into a whole halo at [Fe/H] = −3. If a neutron star merger occurs in a halo with a gas mass of $\sim 10^6$ $M_\odot$, the [Eu/Fe] ratios are expected to be [Eu/Fe] = 1.56. In the case of a halo with a gas mass of $\sim 10^8$ $M_\odot$, the [Eu/Fe] ratio should be −0.34. Results of our simulation support this argument. Our results suggest that r-II stars in the Milky Way halo are formed in halos with a gas mass of $\sim 10^6$ $M_\odot$. These halos can be found at the early stages of Milky Way halo formation or later accreted sub-halos. Roederer et al. [24, 25] have shown that *r*-process enhanced stars may have accreted origin based on kinematics data of these stars in Gaia Data Release 2 [6]. This observation is consistent with our implication.

7.5 Summary

In this chapter, we performed a series of cosmological zoom-in simulations of a galaxy to obtain the self-consistent picture for the formation of the Milky Way halo and satellite dwarf galaxies through the enrichment of heavy elements. We computed the galaxy with a halo mass of 10^{10} $M_\odot$ at redshift 2 to resolve the scale of satellite dwarf galaxies. Most extremely metal-poor stars are formed before 1 billion years from the beginning of the simulation. In the galaxy, there are star-to-star scatters of *r*-process elements in low metallicity. We find that *r*-process rich metal-poor stars are formed in halos with a total mass of $\sim 10^8$ $M_\odot$ and gas mass of 7–9×10^6 $M_\odot$.

References

1. Abbott BP et al (2017a) GW170817: observation of gravitational waves from a binary neutron star inspiral. Phys Rev Lett 119(16):161101
2. Abbott BP et al (2017b) Multi-messenger observations of a binary neutron star merger. Astrophys J Lett 848:L12
3. Bauswein A, Goriely S, Janka H-T (2013) Systematics of dynamical mass ejection, nucleosynthesis, and radioactively powered electromagnetic signals from neutron-star mergers. Astrophys J 773:78
4. Beers TC, Christlieb N (2005) The discovery and analysis of very metal-poor stars in the galaxy. Annu Rev Astron Astrophys 43:531–580
5. Behroozi PS, Wechsler RH, Conroy C (2013) The average star formation histories of galaxies in dark matter halos from z = 0 − 8. Astrophys J 770:57
6. Brown AGA et al (2018) Gaia data release 2-summary of the contents and survey properties. Astron Astrophys 616:A1
7. Frebel A, Norris JE (2015) Near-field cosmology with extremely metal-poor stars. Annu Rev Astron Astrophys 53:631–688
8. Freiburghaus C, Rosswog S, Thielemann F-K (1999) R-process in neutron star mergers. Astrophys J Lett 525:L121–L124

9. Gill SPD, Knebe A, Gibson BK (2004) The evolution of substructure—I. A new identification method. Mon Not R Astron Soc 351:399–409
10. Goriely S, Bauswein A, Janka H-T (2011) R-process nucleosynthesis in dynamically ejected matter of neutron star mergers. Astrophys J Lett 738:L32
11. Hahn O, Abel T (2011) Multi-scale initial conditions for cosmological simulations. Mon Not R Astron Soc 415:2101–2121
12. Honda S, Aoki W, Kajino T, Ando H, Beers TC, Izumiura H, Sadakane K, Takada-Hidai M (2004) Spectroscopic studies of extremely metal-poor stars with the subaru high dispersion spectrograph. II. The r-process elements, including thorium. Astrophys J 607:474–498
13. Hopkins PF, Quataert E, Murray N (2011) Self-regulated star formation in galaxies via momentum input from massive stars. Mon Not R Astron Soc 417:950–973
14. Ishimaru Y, Wanajo S, Prantzos N (2015) Neutron star mergers as the origin of r-process elements in the galactic halo based on the sub-halo clustering scenario. Astrophys J Lett 804:L35
15. Ji AP, Frebel A, Chiti A, Simon JD (2016) R-process enrichment from a single event in an ancient dwarf galaxy. Nature 531:610–613
16. Knollmann SR, Knebe A (2009) AHF: Amiga's halo finder. Astrophys J 182:608–624
17. Komiya Y, Shigeyama T (2016) Contribution of neutron star mergers to the r-process chemical evolution in the hierarchical galaxy formation. Astrophys J 830:76
18. Korobkin O, Rosswog S, Arcones A, Winteler C (2012) On the astrophysical robustness of the neutron star merger r-process. Mon Not R Astron Soc 426:1940–1949
19. Naiman JP, Pillepich A, Springel V, Ramirez-Ruiz E, Torrey P, Vogelsberger M, Pakmor R, Nelson D, Marinacci F, Hernquist L, Weinberger R, Genel S (2018) First results from the IllustrisTNG simulations: a tale of two elements-chemical evolution of magnesium and europium. Mon Not R Astron Soc 477:1206–1224
20. Ojima T, Ishimaru Y, ShinyaWanajo NP, François P (2018) Stochastic chemical evolution of galactic subhalos and the origin of r-process elements. Astrophys J 865(87):87
21. Okamoto T, Eke VR, Frenk CS, Jenkins A (2005) Effects of feedback on the morphology of galaxy discs. Mon Not R Astron Soc 363:1299–1314
22. Planck Collaboration (2016) Planck 2015 results-XIII. Cosmological parameters. Astron Astrophys 594:A13
23. Roederer IU, Mateo M, Bailey III JI, Song Y, Bell EF, Crane JD, Loebman S, Nidever DL, Olszewski EW, Shectman SA, Thompson IB, Valluri M, Walker MG (2016) Detailed chemical abundances in the r-process-rich ultra-faint dwarf galaxy reticulum 2. Astron J 151:82
24. Roederer IU, Hattori K, Valluri M (2018) Kinematics of highly r-process-enhanced field stars: evidence for an accretion origin and detection of several groups from disrupted satellites. Astron J 156(179):179
25. Roederer IU, Sakari CM, Placco VM, Beers TC, Ezzeddine R, Frebel A, Hansen TT (2018) The r-process alliance: a comprehensive abundance analysis of HD 222925, a metal-poor star with an extreme r-process enhancement of [Eu/H] = −0.14. Astrophys J 865(129):129
26. Rosswog S, Korobkin O, Arcones A, Thielemann F-K, Piran T (2014) The long-term evolution of neutron star merger remnants—I. The impact of r-process nucleosynthesis. Mon Not R Astron Soc 439:744–756
27. Saitoh TR, Makino J (2012) A natural symmetrization for the plummer potential. New Astron 17:76–81
28. Saitoh TR, Daisaka H, Kokubo E, Makino J, Okamoto T, Tomisaka K, Wada K, Yoshida N (2008) Toward first-principle simulations of galaxy formation: I. How shouldwe choose star-formation criteria in high-resolution simulations of disk galaxies? Publ Astron Soc Jpn 60:667–681
29. Saitoh TR, Daisaka H, Kokubo E, Makino J, Okamoto T, Tomisaka K, Wada K, Yoshida N (2009) Toward first-principle simulations of galaxy formation: II. Shock-induced starburst at a collision interface during the first encounter of interacting galaxies. Publ Astron Soc Jpn 61:481–486
30. Shen S, Cooke RJ, Ramirez-Ruiz E, Madau P, Mayer L, Guedes J (2015) The history of r-process enrichment in the milky way. Astrophys J 807:115

31. Springel V (2005) The cosmological simulation code GADGET-2. Mon Not R Astron Soc 364:1105–1134
32. Tanaka M et al (2017) Kilonova from post-merger ejecta as an optical and near-Infrared counterpart of GW170817. Publ Astron Soc Jpn 69:102
33. Utsumi Y et al (2017) J-GEM observations of an electromagnetic counterpart to the neutron star merger GW170817. Publ Astron Soc Jpn 69:101
34. van de Voort F, Quataert E, Hopkins PF, Kereš D, Faucher-Giguère C-A (2015) Galactic r-process enrichment by neutron star mergers in cosmological simulations of a milky way-mass galaxy. Mon Not R Astron Soc 447:140–148
35. Wanajo S, Sekiguchi Y, Nishimura N, Kiuchi K, Kyutoku K, Shibata M (2014) Production of all the r-process nuclides in the dynamical ejecta of neutron star mergers. Astrophys J Lett 789:L39

Chapter 8
Conclusions and Future Prospects

Abstract In this book, we performed a series of N-body/smoothed particle hydrodynamics simulations of dwarf galaxies to clarify the enrichment of heavy elements. This book aims to understand (1) the contribution of neutron star mergers and supernovae (core-collapse supernovae, hypernovae, electron-capture supernovae, and type Ia supernovae) to the enrichment of heavy elements and (2) the relation between the chemodynamical evolution of galaxies and the abundances of heavy elements. Here we show the new findings in this book.

8.1 Conclusions

The main conclusions can be drawn from Chaps. 4–7. In Chap. 4, we newly put the effect of electron-capture supernovae into a series of chemodynamical simulations. We found that stars which have [Zn/Fe] $\gtrsim 0.5$ reflect the ejecta of electron-capture supernovae. In the early phases, gases which have high [Zn/Fe] ratios from the ejecta of electron-capture supernovae remain because of the inhomogeneity of spatial distribution of metallicity. Scatters seen in the ratios of [Zn/Fe] at higher metallicity ([Fe/H] > -2.5) are due to the contribution of type Ia supernovae. These stars are formed at $\lesssim 4$ billion years from the start of the simulation. The scatters of [Zn/Fe] at high metallicity are the same as [Mg/Fe]. We also found that it is difficult to explain the observed relation of [Zn/Fe] in low metallicity without the contribution from electron-capture supernovae. On the other hand, the observed ratios of [Zn/Fe] can be explained without the contribution of hypernovae.

In Chap. 5, we performed a series of chemodynamical simulations of an isolated dwarf galaxy model assuming that neutron star mergers are the major astrophysical site of the r-process. We found that models with metal mixing in a star-forming region can produce stars with [Eu/Fe] in extremely metal-poor stars. Neutron star mergers with a merger timescale of 100 million years contribute to the enrichment of r-process elements in dwarf spheroidal galaxies. Due to the suppressed star formation

© Springer Nature Singapore Pte Ltd. 2019
Y. Hirai, *Understanding the Enrichment of Heavy Elements*
by the Chemodynamical Evolution Models of Dwarf Galaxies,
Springer Theses, https://doi.org/10.1007/978-981-13-7884-3_8

rate ($\sim 10^{-3}$ $M_{\odot}$ yr^{-1}) in the early phases, [Fe/H] does not increase at $\lesssim 300$ million years in our model. This result suggests that neutron star mergers with timescales of $\lesssim 300$ million years can contribute to the enrichment of r-process elements in dwarf galaxies. We also found that neutron star mergers with a rate of ~ 1000 Gpc^{-3}yr^{-1} are enough to explain the observed [Eu/Fe] ratio if they produce $\sim 10^{-2}$ $M_{\odot}$ of r-process elements in one event. These rate and yield are consistent with the estimation from the observation of the gravitational waves from a neutron star merger [1, 14].

We also discussed the relationship among the structure of halos, star formation histories, and enrichment of r-process elements. To produce galaxies with different star formation histories and stellar mass, we changed the central density and total mass of halos of our isolated dwarf galaxy model. We found that the distribution of r-process elements in extremely metal-poor stars depends on the star formation rates in the early phases of galaxy evolution. Models with star formation rates less than 10^{-3} $M_{\odot}$yr^{-1} produce stars with r-process elements in [Fe/H] $\lesssim -3$. The star formation rates in the early phase depend on the central density but not on the total mass of halos. We found that the early star formation rates are suppressed to be less than 10^{-3} $M_{\odot}$yr^{-1} in galaxies which have dynamical times of ~ 100 million years. Neutron star mergers with merger timescales from 10 to 500 million years do not affect the result significantly. We do not need neutron star mergers with unlikely short merger timescale to explain the observation of r-process elements if the star formation rate is suppressed to be less than 10^{-3} $M_{\odot}$yr^{-1}.

Chapter 6 has shown that abundances of heavy elements can be used as a tracer of the efficiency of metal mixing in galaxies. We use Mg, Ba, and Zn as a tracer of metal mixing. We found that the scaling factor for metal diffusion should be less than 0.01 to explain the observed Ba abundances in dwarf galaxies. This value is consistent with the estimation of the turbulence theory and experiment. The timescale of metal mixing is estimated to be ~ 40 million years from the gas phase Ba abundances. This timescale is shorter than the dynamical timescales (~ 100 million years) of dwarf galaxies. The scaling factor for metal diffusion from ~ 0.01 can also explain the [Zn/Fe] ratio in metal-poor stars. This value is consistent with the estimation from Ba abundances.

In Chap. 7, we performed a series of high-resolution cosmological zoom-in simulations to connect the enrichment of r-process elements to the galaxy formation. We found that most extremely metal-poor stars are formed before one billion years from the beginning of the simulation. This result implies that extremely metal-poor stars observed in the Local Group galaxies contain information about the enrichment history of elements within one billion years from the formation of first stars. In [Fe/H] < -2, there are star-to-star scatters of [Eu/Fe] ratios. This feature is seen in stars in the Milky Way halo. Scatters of [Eu/Fe] ratios caused by the contribution of accreted dwarf galaxies in the early phases of galaxy formation. We found that the r-process rich metal-poor stars are formed in halos with a gas mass of 7–9 $\times$ 10^6 $M_{\odot}$. This result suggests that r-process rich stars seen in the Milky Way halo may come from the accreted galaxies similar to Reticulum II ultra-faint dwarf galaxy.

Figure 8.1 illustrates the scenario proposed in this book. We found that neutron star mergers can be the major contributor to the enrichment of r-process elements in

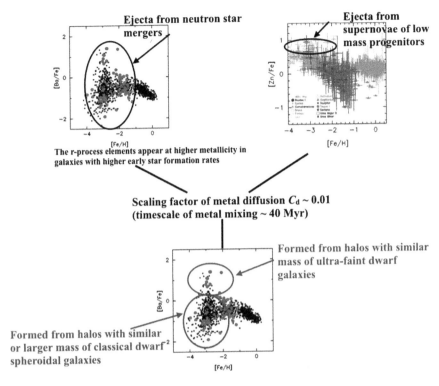

Fig. 8.1 The scenario proposed in this book

galaxies. We also found that electron-capture supernovae contribute to the enrichment of Zn in low metallicity. By using the heavy element abundances, we showed that the timescale of metal mixing is ~40 million years in galaxies. We can extend these results to the enrichment of heavy elements in a cosmological context. From the cosmological zoom-in simulations with enrichment of r-process elements, we propose the scenario that r-process rich stars in the Milky Way halo are formed in small halos with the size comparable to those of the present ultra-faint dwarf galaxies. This result demonstrates that abundances of heavy elements in Local Group galaxies can be an excellent tracer of the early phases of evolutionary histories of Local Group galaxies.

8.2 Future Prospects

In this book, we performed high-resolution chemodynamical simulations of dwarf galaxies newly including the effects of neutron star mergers and electron-capture supernovae. For neutron star mergers, we assume the rates within the range of estimation from observed binary pulsars (Chaps. 5, 6, and 7). This rate is still very uncertain even if there is an observation of gravitational waves from a neutron star merger [1]. In Chap. 4, we studied the enrichment of Zn, but we cannot fully explain the decreasing trend of [Zn/Fe] ratios toward higher metallicity. Also, even if we performed one of the highest resolution simulations, we are still unable to resolve each star in the simulation. These problems will be resolved shortly.

8.2.1 Multi-messenger Astronomy

Enrichment of r-process elements from neutron star mergers will be able to study using the multi-wavelength observations and detailed simulations. The third observation run (O3) will be operated around 2019 in the gravitational wave detector, advanced LIGO, and Virgo. The other detector, KAGRA will start operation at around 2020. Full operation of three detectors will be started around 2022. In the full operation, we expect to detect neutron star mergers ~ 10 events per year. These observations will improve the estimation of the rate of neutron star mergers. These observations will also be able to constrain the yields, delay time distributions, and neutron star kick distributions.

Estimation of rates and kick distributions is critical to understand the enrichment of r-process elements fully. Neutron stars receive kick when the supernova explosion. Binary neutron stars with kick velocities more enormous than the escape velocity of the host galaxy will merge outside of the galaxy. Safarzadeh and Cote [9] showed that $\sim 40\%$ of all formed binary neutron stars do not contribute to the enrichment of r-process elements. These escaped binary neutron stars do not contribute to the enrichment of r-process elements inside the galaxy but distribute elements to the intergalactic medium.

In the Milky Way halo, stars with [Fe/H] $\lesssim -3.5$ have low-level Ba abundances ([Ba/Fe] ~ -1.5). Ultra-faint dwarf galaxies except for Reticulum II also have depleted abundances of r-process elements [3]. Origin of these stars is not yet understood. Materials ejected to the intergalactic medium would account for forming such stars. We expect that future chemodynamical simulations including the effects of neutron star kick resolve these issues.

8.2.2 Star-By-Star Simulations

It is our quest to understand the evolutionary histories of galaxies from the formation of first stars to the present. Understanding small systems such as ultra-faint dwarf galaxies and globular clusters is the key to clarify the evolutionary histories of the Milky Way halo. These objects contain information about the early Universe. To understand the formation and evolution of these systems, we need to perform a series of very high-resolution simulations that can resolve each star (star-by-star simulations). These simulations will be able to perform using the future upgraded supercomputing systems and improved computational codes.

Recent surveys discovered many ultra-faint dwarf galaxies in the Local Group (e.g., Simon and Geha [11]). Weisz et al. [15] showed that star formation stops earlier in lower mass galaxies. Supernova feedback is effective in these galaxies [7]. By analyzing star formation histories of ultra-faint dwarf galaxies, it is possible to constrain the efficiency of supernova feedback, which is not well understood.

Formation of globular clusters is one of the open questions in astronomy. Globular clusters are one of the main components in the Local Group. They consist of old stars with ages over 10 billion years. There are star-to-star scatters of r-process elements, but no scatters of Fe in the globular cluster, M15 [8, 12]. This observation points to the astrophysical sites of r-process elements and evolutionary histories of globular clusters. We expect that future simulations that can resolve each star will be able to clarify the formation and evolution of globular clusters.

8.2.3 Connection Between Near and Far Fields

Our simulations can apply to distant and larger objects. Understanding the enrichment of Zn in galaxies will clarify the chemical evolution in the Universe. Zn has been used for the indicator of metallicity in damped Lyα systems. The increasing trend of [Zn/Fe] ratios toward lower metallicity would be resolved in star-by-star simulations that can reflect the yields of each supernova. Chemical evolution in the high redshift universe will be clarified by the simulations of damped Lyα systems including enrichment of Zn from supernovae.

The cosmological zoom-in simulations discussed in Chap. 7 can apply for clusters of galaxies. X-ray observations of Perseus clusters showed that the ratio of iron-peak elements and Fe is consistent with those of solar system [6]. The metallicity distributions in the intergalactic medium are uniform in the outskirts of Perseus and Virgo clusters [10, 16]. These observations give us insight into the metal mixing in the intergalactic medium. Future simulations of a cluster of galaxies will probe the metal mixing in a larger scale.

Future observations will give us the data that can directly compare with the high-resolution simulations. The Gaia mission will provide astrometry and photometry data over one billion stars down to the magnitude of $\sim$20 [2, 4, 5]. The Subaru prime focus spectrograph will give us chemodynamical data of the Local Group galaxies [13]. The next generation large telescopes such as the Thirty Meter Telescope (TMT), the Giant Magellan Telescope (GMT), and the European Extremely Large Telescope (E-ELT) will observe the Universe down to the era of the formation of first galaxies. The comparison between future simulations and observations will lead us to understand the formation and evolution of galaxies.

References

1. Abbott BP et al (2017) GW170817: observation of gravitational waves from a binary neutron star inspiral. Phys Rev Lett 119(16):161101
2. Brown AGA, et al (2018) Gaia data release 2. Summary of the contents and survey properties. Astron Astrophys 616, A1, A1
3. Frebel A, Norris JE (2015) Near-field cosmology with extremely metal-poor stars. Annu Rev Astron Astrophys 53:631–688
4. Gaia Collaboration (2016a). Gaia Data Release 1. Summary of the astrometric, photometric, and survey properties. Astron Astrophys 595, A2
5. Gaia Collaboration (2016b) The Gaia mission. Astron Astrophys 595:A1
6. Hitomi Collaboration et al (2017) Solar abundance ratios of the iron-peak elements in the Perseus cluster. Nature 551:478–480
7. Okamoto T, Frenk CS, Jenkins A, Theuns T (2010) The properties of satellite galaxies in simulations of galaxy formation. Mon Not R Astron Soc 406:208–222
8. Otsuki K, Honda S, Aoki W, Kajino T, Mathews GJ (2006) Neutron-capture elements in the metal-poor globular cluster M15. Astrophys J Lett 641:L117–L120
9. Safarzadeh M, Cote B (2017) On the impact of neutron star binaries' natalkick distribution on the Galactic r-process enrichment. Mon Not R Astron Soc 471:4488–4493
10. Simionescu A, Werner N, Urban O, Allen SW, Ichinohe Y, Zhuravleva I (2015) A uniform contribution of core-collapse and type Ia supernovae to the chemical enrichment pattern in the outskirts of the virgo cluster. Astrophys J Lett 811:L25
11. Simon JD, Geha M (2007) The kinematics of the ultra-faint milky way satellites: solving the missing satellite problem. Astrophys J Lett 670:313–331
12. Sneden C, Kraft RP, Shetrone MD, Smith GH, Langer GE, Prosser CF (1997) Star-To-Star abundance variations among bright giants in the metal-poor globular cluster M15. Astrophys J Lett 114:1964
13. Takada M, Ellis RS, Chiba M, Greene JE, Aihara H, Arimoto N, Bundy K, Cohen J, Dore O, Graves G, Gunn JE, Heckman T, Hirata CM, Ho P, Kneib J-P, Le F'evre O, Lin L, More S, Murayama H, Nagao T, Ouchi M, Seiffert M, Silverman JD, Sodre L, Spergel DN, Strauss MA, Sugai H, Suto Y, Takami H, Wyse R (2014) Extragalactic science, cosmology, and Galactic archaeology with the Subaru Prime Focus Spectrograph. Publ Astron Soc Jpn 66:R1
14. Tanaka M et al (2017) Kilonova from post-merger ejecta as an optical and near- Infrared counterpart of GW170817. Publ Astron Soc Jpn 69:102
15. Weisz DR, Dolphin AE, Skillman ED, Holtzman J, Gilbert KM, Dalcanton JJ, Williams BF (2014) The star formation histories of local group dwarf galaxies. I. Hubble space telescope/wide field planetary camera 2 observations. Astrophys J 789, 147
16. Werner N, Urban O, Simionescu A, Allen SW (2013) A uniform metal distribution in the intergalactic medium of the Perseus cluster of galaxies. Nature 502:656–658

About the Author

Yutaka Hirai

Address: RIKEN Center for Computational Science, 7-1-26 Minatojima-minami-machi, Chuo-ku, Kobe, Hyogo 650-0047, Japan
Tel: +81-78-940-5816
Fax: +81-78-304-4971
E-Mail: yutaka.hirai@riken.jp
Web: http://th.nao.ac.jp/MEMBER/hirai/index.html

Appointments

- Special Postdoctoral Researcher (SPDR), RIKEN (Apr 2018–)
- Research Fellow of Japan Society for the Promotion of Science (JSPS Research Fellow), The University of Tokyo (Apr 2015–Mar 2018)

Education

- Doctor of Philosophy (Ph.D.) in Astronomy, The University of Tokyo (2015–2018)
- Master of Science (MSc) in Astronomy, The University of Tokyo (2013–2015)
- Bachelor of Science (BSc) in Physics, Waseda University (2009–2013)

Service

- Reviewer for The Astrophysical Journal, Monthly Notices of the Royal Astronomical Society, and JPS Conference Proceedings

Teaching

- Part-time lecturer at Konan University (2018–)

Research Statement

My research goal is to understand the formation and evolution of galaxies comprehensively. I am particularly interested in the enrichment of heavy elements in

© Springer Nature Singapore Pte Ltd. 2019
Y. Hirai, *Understanding the Enrichment of Heavy Elements*
by the Chemodynamical Evolution Models of Dwarf Galaxies,
Springer Theses, https://doi.org/10.1007/978-981-13-7884-3

galaxies. At the time of the Big Bang, the Universe consists of hydrogen, helium, and a small amount of lithium. However, we are now living in the Universe with more than 90 kinds of elements. Most of these elements are synthesized in stars and distributed to the space associated with the death of a star. The next generation stars inherit the abundances of elements previously distributed to the Universe. Understanding the heavy elements in galaxies, therefore, directly lead us to clarify the evolutionary history of the Universe. My approach is to conduct the hydrodynamic simulations of galaxies. I try to connect imprint shown in the observed galaxies and their evolutionary histories. Through these efforts, we can obtain a comprehensive understanding of the evolutionary histories of the Universe from the Big Bang to the present.

Index

© Springer Nature Singapore Pte Ltd. 2019
Y. Hirai, *Understanding the Enrichment of Heavy Elements
by the Chemodynamical Evolution Models of Dwarf Galaxies*,
Springer Theses, https://doi.org/10.1007/978-981-13-7884-3

133

CPSIA information can be obtained
at www.ICGtesting.com
Printed in the USA
LVHW051937160519
618116LV00002B/10/P